Science and Empire in the Nineteenth Century

Science and Empire in the Nineteenth Century: A Journey of Imperial Conquest and Scientific Progress

Edited by

Catherine Delmas, Christine Vandamme and Donna Spalding Andréolle

CAMBRIDGE SCHOLARS

PUBLISHING

Science and Empire in the Nineteenth Century:
A Journey of Imperial Conquest and Scientific Progress,
Edited by Catherine Delmas, Christine Vandamme and Donna Spalding Andréolle

This book first published 2010

Cambridge Scholars Publishing

12 Back Chapman Street, Newcastle upon Tyne, NE6 2XX, UK

British Library Cataloguing in Publication Data
A catalogue record for this book is available from the British Library

ISBN (10): 1-4438-2559-X, ISBN (13): 978-1-4438-2559-7

TABLE OF CONTENTS

INTRODUCTION

If science and technology contributed to the development of European civilisation, they also facilitated the exploration of the world and the expansion of territories. The word "civilization," introduced into the French and English languages in the eighteenth century, implies the idea of social, political, economic progress, but also dovetails with expansion and colonisation. Civilisation, from the Latin *civis*, or *civilitas*, combines two notions: the state of being civilised, and the process or action of civilising others, by exporting a model, be it a model of government or a cultural, religious and social model. In his introduction to *The Decline of the West* (1952), Oswald Spengler, basing his analysis on the historical evolution from Greek culture to Roman civilisation, explains that civilisation is synonymous with achievement and money interests, and most importantly that imperialism is its logical conclusion.

The extraordinary development of science from the sixteenth century onwards through the Enlightenment and the industrial Revolution and its decisive economic, ideological and political impact throughout the nineteenth century marked a new stage in the progress of western civilisation, often opposed in the Victorian era to a previous, more primitive stage of development, the dark ages of "barbarism," which could consequently be enlightened thanks to the civilising mission undertaken by the western world. Science and technology were both the means which made this "civilizing mission" possible, and the proclaimed goal of discovering new peoples, species and territories that contributed to the further advancement of knowledge. Confronted with the unknown, scientists collected data, examined and classified plant, animal and sometimes human species to encompass the multiplicity and diversity of the earth. Nineteenth century scientists were interested in an extensive range of research fields—botany, anthropology, hydrography, map-making or geology to quote but a few.

The issue at stake in this volume is the role of science as a way to fulfil a quest for knowledge, a tool in the exploration of foreign lands, a central paradigm in the discourse on and representations of otherness. The interweaving of scientific and ideological discourses is not limited to the geopolitical frame of the British empire in the nineteenth and early twentieth centuries but extends to the rise of the American empire as well.

The fields of research tackled in this book are human and social sciences (anthropology, ethnography, cartography, phrenology), which thrived during the period of imperial expansion, racial theories couched in pseudo-scientific discourse, natural sciences, as they are presented in specialised or popularised works, in the press, in travel narratives—at the crossroads of science and literature—, in essays, but also in literary texts.

Such approaches allow for the analysis of the link between knowledge and power as well as of the paradox of a scientific discourse which claims to seek the truth while at the same time both masking and revealing the political and economic stakes of Anglo-saxon imperialism. The analysis of various types of discourse and/or representation highlights the tension between science and ideology, between scientific "objectivity" and propaganda, and stresses the limits of an imperialist epistemology which has sometimes been questioned in more ambiguous or subversive texts.

The scientific discoveries of the nineteenth century and the epistemological crisis at the turn of the twentieth century also often triggered existential disquiet and anguish, metaphysical questioning, which found a convenient outlet in a quest for origins and myths, a fantasised return to a pre-industrial state and an idealisation of nature as well as the conquest or imaginary representation of newly explored countries. Science can thus engender or reveal two opposed visions of the world: a reassuring one which presents a well-ordered world with clear limits and a frightening one which features a complex and boundless universe which escapes the control of science and imperialism.

This book examines such issues as the plurality of scientific discourses, their historicity, the alienating dangers of reduction, fragmentation and reification of the Other, the interaction between scientific discourse and literary discourse, the way certain texts use scientific discourse to serve their imperialist views or, conversely, deconstruct and question them.

The first chapter shows that scientific activities are often directly or indirectly linked with military intelligence and political interests, and analyses how structures of power underlie cartography and narratives of exploration—travelogues and reports. Interest in archeology, epigraphy, geology and hydrography led Charles Doughty to explore the desert of Arabia in quest of Nabatean monuments and inscriptions, and to draw a map of the peninsula. His map and the data collected in his narrative were later used by T.E. Lawrence when he became the leader of the Arab Revolt, thus showing the link between cartography and geostrategy, knowledge and power as Catherine Delmas explains. Work for the secret services was often conducted under the cover of scientific research—archeology and topography—as the examples of Gertrude Bell in Syria and T.E. Lawrence in Sinai and at Karkemish testify. Cartography undertaken by the Western Survey expedition and made possible by the Palestine Exploration Fund combined a double interest in Biblical land and contemporary Palestine, religious and political issues, ancient history and geostrategic interests at the time of the Ottoman empire, as shown by the bond between the PEF and the War Office at the end of the 19th century. A map can thus be ordered, funded and appropriated by various circles, be they scientific, political or religious, and invites a "multiplicity of readings" revealing structures of power and conflicting interests as Stéphanie Prévost shows.

The mapping of Ireland by the Ordnance Survey which began in 1820 also testifies to the collusion of science and colonial power, not only in the use made of the map—if we consider its administrative, economic and social functions, census, taxation and colonial policy—but in the rhetoric which underlies it and which reveals norms and values as Valérie Morisson explains. For J. B. Harley, a map is a text, in the broad sense of the word, a system and an "instrument of persuasion." It relies on selection, classification, interpretation, a choice of symbols and as such it is a construct which reveals its author's scientific and cultural background. Mapping obeys certain rules which are scientific but are also "related to values, such as those of ethnicity, politics, religion or social class, and they are also embedded in the map-producing society at large, and in its other forms of representation." (Harley, in Barnes and Duncan 1992, 236). Maps are a mode of representation, anchored in culture and historicity; they reveal a point of view, a way of apprehending the world, and structures of power. They can lie, distort and manipulate, and they can be used for propaganda or to spread ideology as Monmonnier explains in *How to Lie with Maps* (1991). They respond to the need for control and surveillance in a colonial context, which they help to implement by "legitimizing territorial conquest, economic exploitation and cultural imperialism" (Monmonier, 90). For Tiffin and Lawson in *De-scribing Empire. Post colonialism and Textuality,* "maps are productions of complex social forces; they create and manipulate reality as much as they record it" (1994, 116). Even the representation of blanks on a map indicates gaps in western knowledge as well as a "process of erasure of existing social and geo-cultural formations in preparation for the projection and subsequent emplacement of a new order" (Tiffin and Lawson, 116). Cartography is thus at the cusp between science and the rhetoric of power.

Cartography and narratives of exploration often reveal nationalist discourse and imperialist ideology, concealed, in Doughty's case, behind an overt interest in man's origins and the desire to preserve the purity of the English language in a travelbook which interlaces scientific and literary discourse. Catherine Delmas thus shows that the purpose of his journey of exploration was to serve and improve European knowledge through the data he had collected, and to defend the faith of a Christian. Doughty's study and practice of natural or "Liberal sciences," as he called them, cast light on the multifaceted activities of a scientist in the nineteenth century; they also reveal his ideological and scientific Eurocentric point of view, based on analogy, classification, typology, i.e. a European point of reference and system of thought which is "arborescent" according to Deleuze and Guattari in *A Thousand Plateaus* (5) and fails to account for the multiplicity, hybridity and heterogeneity of the world. Although a map, contrary to tracing, is "open and connectable in all its dimensions" (Deleuze and Guattari 1980, 13) and has, like a rhizome, "multiple entryways" inviting to nomadology, cartography in the nineteenth century *did* illustrate colonial powers' determination to appropriate and control territories, to turn nomadic space into structured and administered territories. As Edward Casey explains, "[n]owhere is Eurocentrism—that most insidious long-lived form of ethnocentrism—more manifest than in the case of Western cartography" (Casey 2002, 194).

Cartography and the memoirs which were published with the O.S. maps of Ireland had, however, a paradoxical effect as Valérie Morisson shows. The renaming of Irish places into English and the record of Irish names before they disappeared aroused a new interest in toponymy; the memoirs which collected data in "archeology, spelling, genealogy and philology" led to a national and regional awareness, "which would fuel the forging of a strong national identity during the Celtic Revival." Cartography, which shows a "dichotomy between the scholarly observer and the objectified, uncivilized other" (Valérie Morisson), can thus have unexpected effects when it strengthens local nationalism. Instead of illustrating the rhetoric of colonial power, it may become an instrument of resistance which shows that structures of power can be double-edged.

But such a paradoxical side effect of imperial *surveillance* resulting in the strengthening of the colonised subject's position, instead of its intended subjection, remains anecdotal and relatively exceptional. In the second chapter of the book, a close look at the way science or pseudo-scientific theories were used in the nineteenth century to classify both plant and human specimens testifies to the inextricable link between science, empire and the subjugation of the *other*. In her book on *Imperial Eyes, Travel Writing and Transculturation* (1992), Marie Louise Pratt explains how the rise of natural sciences in the eighteenth century, based on specimen gathering, naming, collecting, and a classificatory system devised by Linnaeus, contributed to "othering" newly discovered people (30). Classification and nomenclature are nowhere more central than with botany; and it is very illuminating to note that the binomial nomenclature set up by Linnaeus is still in use nowadays even though his method of classification was criticised as early as the beginning of the nineteenth century for its reductiveness and relative arbitrariness. The taxonomy was too much based on distinguishing features between plants—rather than their similarities—and the selected criteria were exclusively morphological (especially the various types of sexual organs); more essential criteria having to do with the fundamental physiological processes of reproduction and nutrition were simply ignored. Similarly, concerning the classification of human specimens, even though the late nineteenth century saw the development of what was to become modern anthropology, there was a tendency to consider different "races" as distinct in their morphological characteristics (the colour of the skin or the size of the skull for instance) rather than reflect on what they had in common, physiologically or culturally. In Australia for instance, there was a temptation throughout the nineteenth century to use the scientific idea of classification as a model of social classification, assigning a place to both coloniser and colonised. Christine Vandamme shows such an indeniable link between the idea of colonisation, cultivation and classification as a powerful nexus for both scientific and social control. David Malouf's *Remembering Babylon* (1994) remains indeed a literary cornerstone in the examination of the entanglement between a scientific approach—here botanical, in the manner of famous botanist Joseph Banks—and imperial interests. Native plant specimens were forced into a systematic grid imported from the West and if they did not contribute to the general dictate of "improvement," it was simply as if they never existed and they were ignored. In an almost identical way, Australian native "specimens," Aborigines, were offered specific places in society where they could be put to good use both for the community and supposedly for themselves, with the setting up of successive Aboriginal protection boards all over Australia as early as the 1860s (the first one was created in 1860 in Victoria), the idea being that they had to be placed in the right institutions to be given the most appropriate type of occupation later in life. Classification is thus clearly associated with the typically utilitarian doctrine of improvement and the subjection of science to empire. Such a collusion between science, empire and control was also blatant in the thriving scientific theories revolving on the notions of evolution, races and their respective destinies, which were so prominent at the end of the nineteenth century.

An interesting case in point is the imperial use of medicine to justify theories of racial Darwinism which developed in parallel with imperialist expansion all along the century. In the case of Australian settlement, Anne Le Guellec also underlines the importance of British empiricism as contrasted with both Continental rationalism and a desire, inherited from the Enlightenment, to contribute to the advancement of Western knowledge of diseases and remedies; the encyclopaedic approach with a view to set up a catalogue that would be as comprehensive and exhaustive as possible is not essential here. What is more compelling for British and American colonial medical science throughout the century is to answer the prerequisites of an ideologically loaded agenda, namely proving scientifically the inferiority of colonised races and their greater frailty when faced with diseases. But, concerning Australia, another aspect consisted in warning the nation against the possible damaging effects which colonisation could have on the British settlers themselves in weakening the strength of the white race when transplanted in an environment where adaptation could only lead to degeneration. In her article on the experimental aspects of medical science in Australia, Anne Le Guellec thus shows how scientific discourse could legitimise colonisation first as an experiment in survival and adaptation with the risk of degeneration and finally as an experiment in countering the negative effects of the climate and geography through assimilation of the already dying-out Aboriginal race. Such shifts in the experimental paradigm are typical of the representation and the scientific approach of the Other in an imperial context. As Patrick Brantlinger shows in his book *Rule of Darkness* (2003), at the end of the eighteenth century and in the first half of the nineteenth century, the Enlightenment and Romanticism tended to see the native as endowed with a primitive purity and innocence that could save Western civilisations on the decline, but conversely, in the second half of the century, with the sudden explosion of the colonial expansion, the native was turned into the degenerate savage Great Britain had to save from his barbaric ways (178-9).

Thus, in each period, whether Romantic or mid and late-Victorian, the representation of the Other, whether philosophical, literary or scientific, is a direct result of the prevailing ideology of the time. With the transition from a utilitarian advocacy of freedom, *laissez-faire* and abolitionism which was a direct result of Great Britain's unrivalled industrial and naval supremacy, to a context of fierce economic and geopolitical competition, especially in the colonial world, an increased interest in what the colony could bring to the metropolis became omnipresent and the luxury of furthering scientific knowledge for the sheer sake of universal progress had come to be considered more or less obsolete or amateurish.

Towards the end of the century, science was increasingly used in the service of empire but the reverse statement was rarely true. In Australia for instance, whether with botany and its instrumentalisation as a perfect means of forcing beings, be they plant or human specimens, into a preestablished grid imposed by empire and set up to serve its interests, or with medicine and its successive pseudo scientific experiments in matters of adaptation (or the impossibility thereof), degeneration or even assimilation, science came in handy as a perfect tool for colonial ideology. Both its apparently rigorous and rational systematicity and its more tentative experimental aspects could serve the imperial enterprise of classifying but also of correcting the natural course of evolution and selection by either speeding up the natural disappearance of a doomed race (imported diseases and their deathly toll only proved the inherent inadaptability of the native) or preventing it by a progressive "breeding out" of the Aborigines. As Sheila Whittick points out, the dynamics of classification found a perfect if grim illustration in the blind worship of scientific measurements and the fetishistic collection of human remains that characterised racial science in the second half of the nineteenth century and the beginning of the twentieth century. Indeed, the indigenous body could provide the ultimate tangible proof of the inferiority of the native on a global scale of evolution and thus confirm the pseudo scientific prediction of the necessary and ineluctable "passing of the aborigine" as amateur and self-proclaimed anthropologist Daisy Bates would have it in the title of her famous book published in 1938. In her analysis of the complex entanglement between hard sciences such as anatomy, craniology, phrenology and social sciences like anthropology or ethnography increasingly developping, Sheila Whittick shows how irremediably ideological scientific practice could be. Instead of applying one of the most basic principles of the discipline, namely the law of deduction, imperial anthropology, ethnography and museumization tended to first set up an *a priori* theory of extinction and then try and find scientific data to prove the validity of the theory. In so doing, as Patrick Brantlinger shows in his book *Dark Vanishings*, scientific and imperial discourses coalesce in an epistemologically specious way; he thus terms "extinction discourse" a "discursive formation" (in Foucault's sense) which "does not respect the boundaries of disciplines or the cultural hierarchies of high and low" in presenting side by side the analyses of "humanitarians, missionaries, scientists, government officials, explorers, colonists, soldiers, journalists, novelists and poets" (2003, 1).

In the last section of the chapter, Donna Andréolle and Susan Berthier-Foglar demonstrate how, even in "purely" scientific works such as Samuel George Morton's *Crania Americana* (1839) or Josiah Nott and Gliddon's *Types of Mankind* (1855) race theories prevail. And it is therefore interesting to replace such discursive formations in their respective contexts, namely the first half of the nineteenth century in America, with a continued expansion west and the divisive issue of slavery. Morton tried to be as neutral as possible in presenting some objective data, even though his selection of criteria for assessment can be criticised; some of his conclusions, however, would later on be pounced upon to justify slavery–his claim, for instance, that the black man would usually accept relatively easily a state of affairs detrimental to him and not of his own choosing. Conversely, Morton insisted on the fact that different types of black people could range from "intelligent" to "stupid" and such an argument would most probably have been quickly forgotten. The impact of the political and ideological context was even more striking in the work of his successors Nott and Gliddon who selected the aspects from Morton's work that enabled them to support the theory of polygenism and therefore of the distinct nature of the black man as perfect for enslavement contrary to the Indian or other races. The striking irony of such anthropological research was that, *in fine*, neither the North nor the South really used such scientific theories to defend their respective positions contrary to what was happening only shortly afterwards in Europe with the fierce scramble for new colonies in the period that came to be called new imperialism.

On the North American continent, the political and philosophical approaches to science had another essential concern from the very beginning of the nineteenth century, that of America's *manifest destiny* where the principles inherited from the Enlightenment could best flourish. The expansion to the west of the American empire was simply synonymous with the advance of scientific progress and universal knowledge; it is only in the second half of the century that science became so central to political discourse in trying to assign a respective place not only to Indian natives and black slaves but also to the American empire as opposed to all other empires, British or otherwise. Although attitudes toward empire-building and the role of science therein were obviously inherited from the traditions of the British empire of which they had been an integral part during the colonial period, Americans' commitment to using science in the name of colonisation was seen as a necessity if the

United States were to be "put on the map," so to speak, as the American republic emerged:

> The Americans living in this period of exploding scientific inquiry, the fundamental fact conditioning every thought and deed was the consciousness that they were now an independent nation. With respect to science this meant two things: as the example par excellence of useful knowledge, science must be cultivated to promote the interests, prosperity, and power of the rising American nation; and as the supreme example of the powers of the human mind, the success of science challenged Americans to prove to the world that republican institutions were as favorable to intellectual achievement as they were to liberty. (Greene 1984, 5-6)

The key idea here is the notion of *useful* knowledge, that is the promotion of science through which "agriculture is improved, trade enlarged, the arts of living made more easy and more comfortable, and, of course, the increase and happiness of mankind promoted" (*American Philosophical Society* 1771 in Greene 1984, 6). And as both Mehdi Achouche and Jean-Marie Ruiz point out in their respective articles, at the heart of the American vision of science and the building of the American empire resided the figure of Thomas Jefferson. One of the founders of the American republic, astute politician, inventor, and third president of the United States, Jefferson himself was fascinated with various fields of science—ethnography, meteorology, botany and paleontology to name but a few—as is attested in his *Notes on the State of Virginia* (1781) for example; and he spent much time and energy disproving Comte de Buffon's theories on the supposed degenerating nature of the North American climate on its fauna and flora.

Strangely enough, however, Jefferson was antipathetic to scientific theory as well as uninterested in nomenclature and classification which were in vogue at the turn of the nineteenth century, assuming a purely utilitarian vision of science; thus he found botany to be the most important science simply because plants supplied

> the principal subsistence of life to man and beast, delicious varieties for our tables, refreshments from our orchards, the adornments of our flower-borders, shade and perfume for our groves, materials for our buildings, or medicaments for our bodies (qtd. in Greene 1984, 33)

John C. Greene (in *American Science in the Age of Jefferson*, 1984) concludes, then, that although Jefferson was "unimpressive as a scientist and unimaginative in his attitude toward innovation in science" (33), he was an untiring *promoter* of science. This is most obvious in the numerous scientific expeditions which he organized to explore the territories acquired by the Louisiana Purchase of 1803, the most famous of which was the Lewis and Clark Corps of Discovery expedition of 1804-1806. True to Jefferson's pragmatic approach to science, Meriwether Lewis, captain of the Corps and former personal secretary to Jefferson, was trained by the best scientists of the day in field medicine, ethnography, mineralogy, botany and zoology; in the famous letter addressed to him by Jefferson, he was instructed to map all things of interest, provide detailed information on Indian tribes encountered as well as "other objects worthy of notice" such as

> the soil & face of the country, it's [sic] growth & vegetable productions, especially those not of the U.S;
> the animals of the country generally, & especially those not known in the U.S.;
> the remains or accounts of any which may be deemed rare or extinct;
> the mineral productions of every kind [...];
> climate, as characterized by the thermometer [...] the dates at which particular plants put forth or lose their flower, or leaf, times of appearance of particular birds, reptiles of insects. (qtd. in Ronda 1998, 34-35)

The expedition was to return with a total of over five thousand pages of journals, botanic samples, sketches of animals and Indian tribes and the first accurate maps which banished forever the myth of the Northwest passage; the great irony of this scientific episode of the early American empire is that this work was to fall into oblivion, filed in the archives of the American Philosophical Society and only "rediscovered" at the turn of the twentieth century for the centennial celebration of the Corps of Discovery.

Science was used in other ways to shape empire-building. As Mark Meigs points out, for example, the unearthing and displaying of dinosaur remains in museums such as Charles Peale's in Philadelphia were used to build a popular image of American domination of science. Yet another aspect developed in Jean-Marie Ruiz's article is American fascination for the astronomical model of the planets revolving around the sun which gave birth to the notion of "political gravitation," used to serve expansionism beyond American shores in the late nineteenth century. Feeding into this newer, imperialistic vision was the shift from the Enlightenment view of a rational, stable universe to the Darwinian view of progress as a result of inequality in a world of change and competition for resources. This would not prevent, however, a certain nostalgic return to the veneration of nature as embodied in John Muir in the early twentieth century, and the desire to preserve the receding wilderness in the name of the same science which had been so instrumental in conquering the great American continent.

PART I:

SPACE AND STRUCTURES OF POWER

CHARLES DOUGHTY'S QUEST
AND CRUSADE IN *ARABIA DESERTA*

CATHERINE DELMAS

In his book *The Penetration of Arabia* (1904), David George Hogarth shows that the European knowledge of Arabia in the early 20th century was based on the various journeys of exploration undertaken by Westerners since Ludovico di Varthema in the 16th century and the data they had collected. Arabia at the beginning of the 20th century still challenged European knowledge and imagination as numerous blanks on the map remained to be filled. The country resisted exploration for religious reasons, as European visitors were not allowed to enter or even come near the holy cities of Medina and Mecca. Furthermore, the Arabs mistrusted and were often hostile to Christians. The other reasons were climatic and geological as some extremely dry areas like the Nejd or the Nefud deserts were sparsely populated or uninhabited and made exploration extremely difficult.

Although no explorer of inland Arabia himself, Hogarth, an archeologist and later the curator of the Ashmolean museum in Oxford, derives his presentation of the situation, geology and history of the country from previous accounts, i.e. from European observation and representation, and not from Arab testimonies and direct sources. His book shows that knowledge is a construct, both an archive composed of the various data and documents piling up on Arabia and brought back by European travellers (notes, journals, maps, drawings, as well as stones or artefacts), and an ideogical and cultural construct owing to the selection, analysis and interpretation of documents from a western point of view.

At first the urge to discover the country was motivated by personal reasons, more than political ones: curiosity, the physical challenge a journey through Arabia represented, courage in the face of adversity and danger, fame and recognition which were the rewards for one's efforts, and which resulted in fierce competition between European travellers. They all walked in the steps of their predecessors, but were all eager to be the first explorer to set foot in a so far unknown area and to bring back new information. No European traveller had been able to explore the area of

Meidán Sâlih since Varthema. Ludwig Burckhard went as far as Petra, the Nabatean necropolis, but Charles Doughty, who managed to continue the journey from Petra to Meidán Sâlih, was the first explorer to visit the place and copy the inscriptions on bibulous paper.

When he came back to England, competition with Euting, Burton and Charles Huber, the French explorer, made his recognition difficult. Doughty was an obscure scientist at the time; the British Museum refused to publish the results of his findings and he was acknowledged by the Royal Geographical Society much later, in 1912. As Taylor explains in his biography, Burton was already planning to visit Meidán Sâlih and the Blunts were in Hayil. Some of his discoveries, like the stone he had found in Teyma, were credited to Huber and Euting (Taylor 1999, 225). Some of his results were eventually published in *Globus* in Germany thanks to Alois Sprengler, the author of *Die Alte Geographie Arabiens* (1895), a copy of which he carried with him in Arabia; the inscriptions of Meidán Sâlih were published in France at the Académie des Belles Lettres thanks to Ernest Renan.[1]

The military metaphors used by Hogarth in his book illustrate the fierce competition between the various explorers and cast light on their desire to conquer the unknown—he speaks of scientific "conquest," of "attack" and "penetration;" however they also foreshadow the military penetration of Arabia during the First World War, when Thomas Edward Lawrence, working for the War Office in Cairo, later commissioned by the Arab Bureau directed by Hogarth, led the Arab rebellion and guerilla warfare against the Ottoman occupation. David Hogarth, Thomas Edward Lawrence, Gertrude Bell were archeologists and historians, and they were all present in Palestine and Syria before the war. Lawrence conducted excavations on the archeological site of Karkemish[2] under Hogarth's supervision and met Gertrude Bell there in 1911 (Brown 1991, 36). With Woolley, he explored the Sinai, and the findings were published in a report entitled *The Wilderness of Sin*; meanwhile the officers who accompanied them mapped the area, then controlled by the Turks (Brown 1991, 4). Cartography and the record of topographic data, under the cover of harmless scientific research such as archeology, geology or geography, were obviously useful to the British army and the secret services to which T.E. Lawrence and G. Bell belonged.

However, Charles Doughty (1843-1926), who travelled to the Middle East and explored the Sinai in 1873, hardly corresponds to the portrait of the scientist with covert activities. His biographers underline his "quixotic genius" (Hogarth 1904, 275) and "his almost obsessive search for ancient remains" (Taylor 1999, 54) while one critic regards him as "a kind of allegorical Everyman" and "a Spenserian knight" whose quest was "an antiquarian obsession with the origins of civilisations" (Tabachnick 1981, 48, 94). Both Taylor and Tabachnick insist on his romantic vision of rock formations or ruins. Doughty, who saw himself as a wanderer, a *"saieh,"* "a walker about the world" (Doughty 1979, I 315), seems at first to be disconnected from political interests in the Middle East; he left England in 1870 for a conventional Grand Tour of Europe which took him to Belgium, France, Italy, Malta, Tunisia and Algeria. From 1873 to 1875, he travelled to Greece, Egypt, Palestine and Syria; the desire to explore the Nabatean city of Meidán Sâlih motivated him to go back to the Middle East and after a year in Damascus, where he learnt Arabic and prepared his journey, he explored inland Arabia from 1876 to 1878. Doughty was poor and travelled alone, without the financial support of the Royal Geographical Society and the British Association, and without the recommendations of the British Consul in Damas and of the *Dowla*, the local Turkish authorities. He first accompanied the *Hajj* as far as Meidán Sâlih, then explored the desert, sharing the lives of the Beduins who welcomed him, and stayed in small towns like Teyma, Hayil, Kheybar, Boreyda and Aneyza until he reached Jiddah and left the country.

Doughty was both a poet and a scientist "with a dishevelled mind," said Professor Thomas George Bonney from London University where he studied geology (Taylor 1999, 23). The portrait of the "peripatetic scholar" (Taylor 1999, 27) or mystic wanderer who roamed the Arabian desert in search of Nabatean or Hymiaritic inscriptions, rock formations and ruins seems, at first, to undermine the close link between knowledge and power, science and imperialism drawn by Edward Said in his seminal works *Orientalism* and *Culture and Imperialism*. Doughty's autobiographical narrative, *Travels in Arabia Deserta*, published in two thick volumes ten years after he came back from Arabia, was furthermore poorly received by the critics and did not quite correspond to the readers' and scholars' expectations in matter of travelogues. Yet we may wonder whether, through the study of Doughty's quest and crusade in Arabia, exploration can remain innocent and disconnected from political and military interests, or on the contrary whether scientific contribution was not inevitably linked to imperialism in the Victorian era.

After failing to enter a naval career in keeping with a family tradition of naval officers, squires and clergymen, Doughty studied at Gonville and Caius College, Cambridge. His early interests in fossils and rock formations led him to prefer natural sciences to the Classics and to study geology and archeology. Recent findings in England and France and Lyells' *Principles of Geology* challenged Christian beliefs in divine

Creation in the same way as Darwin's theory of evolution revisited the origins of Man in *The Origins of Species* and *The Descent of Man*. The epistemological break was marked by the rise of archeology, prehistory, geology, i.e. human and natural sciences centered on the origins of man and the earth.

Such was Doughty's quest in Arabia:

> Of surpassing interest to those many minds, which seek after philosophic knowledge and instruction, is the Story of the Earth, Her manifold living creatures, the human generations, and Her ancient rocks.[3]

The emphatic, nearly bombastic tone, the archaic grammatical inversion and use of capital letters cast light on the grandeur of the task undertaken by an enlightened scholar and on the universality of a quest for the origins of time and Creation. His epistemological field is plural, in order to encompass the diversity of the living and ancient world. "Liberal sciences," as he calls them (Doughty 1979, I 643), correspond to human, natural and social sciences, and comprise geology, anthropology, ethnography, archeology, epigraphy, linguistics, natural sciences and medecine, which he practiced with the help of a book, but without any previous formation.

Doughty's travelogue reveals his interest in geology and archeology, two disciplines he constantly associates in order to discover the mystery of Creation. He observes rock formations and sand dunes, the harras or volcanic plateaux, the link between topography and hydrography, and the data he collects are illustrated by numerous drawings. His panoramic gaze, the topographical study combined with the minute observation of rocks, result in a survey of the geomorphology of the country, transcribed onto a map. His archeological study of Graeco-Roman and Nabatean remains likewise encompasses towns, architecture, inscriptions and the stone of Teyma.

The two disciplines mingle when, for example, he points out the volcanic origin of the basalt used to build the ancient city of Umm Jemâl, human traces (a cross and Greek inscriptions) and the Biblical origin— Beth Gamul in Jeremiah—of the town (Doughty 1979, I 50). The adjectives he uses—"eternal," "endless," "indestructible" (I 51)—qualify the basalt itself, and dovetail the scientific approach and the belief in the eternity of divine Creation.

Doughty constantly intertwines geology, archeology and Biblical references when he travels in Palestine, and goes from Damas to Petra and Meidán Sâlih. He quotes the Old Testament to refer to places, to compare the present nomads to the "tent dwellers" of the Bible, their habitat, way of life and activities. His approach is descriptive and comparatist, and Doughty's empiricism, based on observation and deduction, serves to prove the veracity of the Old Testament. The Bible is thus the hypotext on which scientific observation is conducted, and science is the means used to prove the truth of the Biblical text. The reader then wonders whether science is subservient to exegesis, or whether the Bible is the key which is necessary to decipher and interpret the world. The Old Testament is evoked to explain the origin of the desert of Edom and Moab (Doughty 1979, I 83) and conversely geology proves the Bible right: "the high limestone downs and open plains of Ammon, Ruben, Gad and Manasseh …the shallower grounds, we may read in the Hebrew Scriptures, were at all times pastoral." (Doughty 1979, I 56)

It is then possible to draw a metonymic and semiotic correspondence between Doughty's text meant to represent the world and the world as a text, i.e. a collection of signs to be read, deciphered and interpreted, between hermeneutics and exegesis, between Doughty's words reduplicating the world and the Logos at the origin of Creation. Although the questionings or even the anxiety triggered by Lyell's and Darwin's theories are never alluded to, Doughty seems to consider evolution within the frame of divine Creation. He tries to restore a sense of continuity in the fragments and traces he observes in the desert, which is echoed by the narrative itself, a coherent whole written in retrospect and based on impressions and notes jotted down on the way. The correspondence between geology and genealogy, the various geological strata which compose the soil of the desert and the various layers of historical time (Biblical, Himyaritic, Nabatean, pre-Muslim and Muslim periods), illustrated by the inscriptions he finds and copies, also restores continuity as they are all part of a whole. Tabachnik considers that time is, as a result, "suspended" (1981, 87), but it is also paradoxically fragmented (the time of daily experience) and linear, from the alpha to the omega of Creation. By emphasising human, linguistic and temporal continuity, Doughty restores filiation from the Hebrew Prophets and patriarchs, and to the vertical lineage, Doughty also associates horizontal kinship between the various Beduin tribes of the desert.

His quest for the origins of Man and the Earth is thus both scientific and metaphysical, and rooted in the epistemological crisis of the Victorian era. Doughty finds an answer both in the Bible and science which he compares to a temple—"the indestructible temple-building of science, wherein is truth" (Doughty 1979, II 409). Doughty's scientific and religious discourse is, however, only possible in the Christian world; as soon as Islam is evoked, Doughty considers that science is superior and necessary to enlighten people steeped in "superstition" and "fanaticism" (*sic*): "a little salt of science would dissolve all their religion" (1979, I 92).

Travels in Arabia Deserta is the contribution of an individual scientist to the improvement of knowledge, but it is both an autobiographical and anthropological narrative, organised around a central consciousness, mediated by an omnipresent first person narrator and through a unique point of view, fraught with stereotype and prejudice, and focusing on the recreated persona of Khalil, the Arabic transcription of his first name Charles which also means "friend," supposedly perceived through the eyes of the local people. Tension between the universality of Doughty's quest and the subjectivity of observation and of the autobiographical narrative calls into question the objectivity of the scientific report and highlights the relativity of truth.

Furthermore his literary style often overlaps with his scientific discourse; his sublime representation of awe-inspiring volcanic landscapes can verge on the Gothic or lapse into the peacefulness of the pastoral. His modes of representation, which convey affects, go counter to scientific observation and his poetic, alliterative style, often using metaphors and hypallages, is hardly in keeping with the scientific report.

Indeed, Doughty saw himself as a poet more than a scientist or an orientalist as he wrote in a letter to David Hogarth in 1902:

> "In writing the volume *Arabia Deserta*, my main intention [...] was not so much the setting forth of personal wanderings among a people of Biblical interest, as the ideal endeavour to continue the older tradition of Chaucer, Spenser, resisting to my power the decadence of English language (*sic*): so that while my work should be a mere script for Orientalists, it should also be my life's contribution, so far, to literature." (quoted by Taylor 1999, 234).

After *Arabia Deserta,* Doughty persevered in his literary quest by writing an epic poem in 24 cantos and 6 volumes about the origins of the nation, *The Dawn of Britain,* "the worst poetry of the 19th century" said William Blunt (quoted by Taylor 1999, 266). The epic combines his patriotism and the defense of the Mother Tongue. It was followed by "unactable dramas" (Taylor 1999, 268): *Adam Cast Forth, The Cliffs,* and *The Clouds,* and other poems *The Titans* and *Mansoul.* Doughty's obsessional concern was obviously the issue of origins: the origins of Man, of the nation and of the English language, his models being Chaucer and Spenser.

This already appears in *Arabia Deserta* in the archaic vocabulary and syntax, which makes the reading at times difficult and which explains why the book was rejected. A report written in December 1883 by Sir Henry Rawlinson from the Royal Geographical Society deplores both Doughty's style and incompetence as a scientist:

"He had no geographical requirements; no knowledge of instruments; no capacity for useful observation, while on the other hand he seems to have to adopt in the description of this journey, such an extravagant eccentricity of style and language as to make his notes not only unfitted to the pages of a scientific journal, but almost unintelligible to any reader, be he scientific or otherwise […]." (quoted by Taylor 1999, 228)

His quest for the origins of language in the pre-Islamic desert and his interests in epigraphy go hand in hand with his quest for the purity of the English language and the use of archaic grammar, vocabulary and syntax in *Arabia Deserta*. The *hajj* or pilgrimage he joins from Damas to Meidán Sâlih, the tales and conversations heard on the way, the language, are reminiscent of *The Canterbury Tales*.

Doughty's quest and pilgrimage in the company of the *Hajj* also turn into a Quixotic enterprise which conditions discourse and representation. Doughty becomes a crusader, not only in the defense of the English language, but as the herald of the western world and the defender of the British nation and Empire. He constantly opposes the embellished portrait of Khalil, his alter ego and persona carrying the torch of progress, science and Enlightenment, to the Other, whose conversations, fables and proverbs in Arabic are reported in medieval English. Linguistic archaisms used to underline the purity of the English language, are also used to highlight the Other's difference and primitive culture. Hierarchy also applies to the Arabs themselves, and Doughty celebrates the golden age of Arabia Felix, "more civil" (Doughty 1979, I 154) but condemns Islamic Arabia, "a decayed country," "forsaken and desolate," preferring myth to history and turning evolution into decadence.

Doughty's jingoism is blatant when the nostalgia for medieval times reveals the portrait of a Crusader in the holy land, a defender of the Christian faith. Doughty assimilates Islam with superstition and fanaticism, and his personal *jihhad* is to claim and defend his faith among "hostile," "fanatic," "barbarous" (1979, I 73), "savage," "hideous" (I 181), "frenetic, sanguinary" "wretches" (I 251). While Hogarth said he was "a humanitarian agnostic," Tabachnik showed that he followed the example of St Paul and St John, and was influenced by the New Testament more than by the Old Testament; but the tension which opposes the scientist and the Crusader, the quest for the truth and ideology, calls into question his humanitarianism and casts light on his patriotism.

The notions of quest and conquest can hardly be dissociated when Doughty defends British hegemony: "But we by navigation are neighbours to all nations, we encompass the earth with our speech in a moment." (Doughty 1979, II 58). He furthermore contributed to the penetration of Arabia by the British army and by T.E. Lawrence who used his map and his book to lead the Arab revolt. The second volume contains useful information about towns and tribes, clans and territories, kinship, the number of inhabitants and soldiers, the location of wells and oases. Human, geographical and topographical data are the surface level of a scientific report which, according to Lawrence, was "the indispensable foundation of all true understanding of the desert," "the great picture-book of nomad life" but which could be used as a "military text-book [that] helped us to victory in the east."[4]

If Doughty was not a conqueror himself, the quest and crusade undertaken by the poet and scientist in Arabia, his discourse and representations, confirm Edward Said's emphasis on the link between knowledge and power, orientalism and imperialism. Doughty claimed he was not an orientalist according to his own definition of the term, i.e. neither a man telling extravagant stories nor a scholar spending his time in libraries. Yet the plurality of his scientific approach corresponds to the definition given by Edward Said as it encompasses natural, human and social sciences and turns the Other into an object of study. Doughty was not an isolated adventurer or a mad scientist, even with "a dishevelled mind"; he was a man of his time and his affiliation is both cultural and ideological. It is not only blatant in his vindictive tone against Islamic Arabia or in his defense of his faith, language and nation; it also underlies his scientific approach, based on binary oppositions, classification, typology, and hierarchy[5] which connote racial prejudice and denote his hierarchical frame of mind, or "arborescent thought."[6] More than "the herald of the outside world" turned into a legend by the Arabs,[7] he was the herald of the Western world, whose quest and crusade, and epic and scientific discourse, expressed ideological, cultural and religious concerns from a Eurocentric point of view.

Works cited

Doughty, Charles. 1979. *Travels in Arabia Deserta* [1888]. 2 vols. New York: Dover.
—. 1884. *Documents épigraphiques recueillis dans le nord de l'Arabie.* Paris : Imprimerie Nationale.

Brown, Malcolm Ed. 1991. *The Letters of T.E.Lawrence*. London: J.M. Dent and Sons Ltd, 1988. Oxford: Oxford University Press.

Deleuze, Gilles and Félix Guattari. 1980. *Mille Plateaux*. Paris: Les Editions de Minuit.

Delmas, Catherine. 2005. *Ecritures du désert :Voyageurs, romanciers anglophones XIXe-XXe siècles*. Aix en Provence: Presses universitaires de Provence.

Hogarth, David George.1904. *The Penetration of Arabia*. London: Lawrence and Bullen.

—. 1929. *The Life of Charles M. Doughty*. New York: Doubleday.

Lawrence, Thomas Edward.1979. Introduction to *Travels in Arabia Deserta* [1888]. 2 vols. New York: Dover.

Said, Edward W. 1979. *Orientalism*. New York: Vintage Books.

—. 1993. *Culture and Imperialism*. London: Chatto and Windus.

Taylor, Andrew. 1999. *God's Fugitive*. London: Harper Collins.

Tabachnick, Stephen Ely. 1981. *Charles Doughty*. Boston: Twayne Publishers.

"A PERFECT MAP OF PALESTINE" (1872-1880): BIBLICAL GEOGRAPHY, INTELLIGENCE AND PROPHECY

STÉPHANIE PRÉVOST

> This invaluable and enduring work will be the result of the subscriptions of private individuals united by the one common bond of being students of the Bible; *it will be completed without State aid*, and once finished will be a work for all time absolutely indispensable to every future student of the Bible. ('Annual Meeting', *PEF Quarterly Statement*,1875, 110, my italics)

This excerpt from the report on the Western Survey was read by the Archbishop of York at the Annual Meeting of the Palestine Exploration Fund (PEF) held on 10 June, 1875. It points to the objective of the Western Survey expedition, begun in December 1871, which was "the making of a perfect map of Palestine" ("Annual General Meeting," *PEF Quarterly Statement*, 1873, 127). The Western Survey was thus a continuation of the rationale for the establishment of the PEF in 1865, as this line in the prospectus suggests:

> No country should be of so much interest to us as that in which the documents of our Faith were written, and the momentous events they describe enacted. At the same time no country more urgently requires illustration. (Committee of the PEF, 13).

The 1865 prospectus thus invites us to understand the Western Survey map as "dual-purpose historical evidence" (Hopkins 1968, 31), as it was a means to "facilitate the understanding of the land in ancient times," but was also the record of the topography and demographic settlements of contemporary Palestine. Equally importantly, by evoking the funds that made the enterprise financially possible, the Archbishop opposes the generosity of private subscribers to the apparent financial non-commitment of the State. By so doing, he deftly suggests that the Western Survey map may interest others than PEF subscribers, especially the State, but precisely indicts it for not sharing the costs. However, assuming that there was no contact whatsoever at the time of the Western Survey between the PEF and the State would be completely wrong, be it only because the War Office (WO) seconded Royal Engineers to carry out the survey. The report statement undoubtedly needs to be looked into to understand the complex relationship between the WO and the PEF, all the more so as it discloses the multiple uses that the map could eventually know.

The aim of this paper will thus be to look into these potential uses. It will first lead us to consider the relationship between the PEF and the WO and the possible tensions between the two organisations, but mostly their complementarities. After reviewing the material help provided by the State, and more particularly by the WO to the PEF, it will appear that the interest displayed by the WO in the work of the PEF actually corresponded to times of Eastern crises, especially between 1875 and 1878. Eventually, the map, just as a number of PEF or PEF-related publications, will have to be considered from a Biblical geographical standpoint, and more particularly as contributing to reinforce the symbolic power conferred to Palestine—both from a restorationist and from a geostrategic point of view.

The relationship between the PEF and State Departments (Foreign Office and War Office): a gentleman's agreement?

The PEF had undeniably been created in the context of raging controversy about Biblical sites, especially the site of the Holy Sepulchre (Lippman 1988, 46). And yet, it was also heavily dependent on the War Office forces to carry out its expeditions, and that from the start. Indeed, the War Office seconded officers and sappers to the PEF. These had topographical expertise but had also been trained in the military field to gather military intelligence. It was already the case when, between 1867-1870, Charles Wilson of the Royal Engineers completed the Ordnance Survey of Jerusalem for the PEF, but also for the Ordnance Survey, when the department was still part of the War Office.[8] The PEF and the WO were thus interlinked from the early days of the PEF expeditions and would remain so very tightly during the 1870s. But the bond between the structures clearly went beyond staff secondment, as the Western Survey was actually using the same methodology as the Ordnance Survey of England and Wales. The Western Survey map was eventually to become the Ordnance Survey map of Palestine,[9] at the scale of one inch to the mile.

The bond between the PEF and the WO was also reinforced by the fact that the topographical information gathered in the field was regularly sent to the Ordnance Survey Office at Southampton. At the time when publication was considered, in 1877, the collaboration between the WO and the PEF was eventually to express itself in the presentation of the map. Each sheet of the map indeed bears the mention of the PEF at the top and that of "surveyed and drawn under the direction of Lieutenants Conder and Kitchener, R.E." at the bottom (PEF/ MINS/ 5/24/1878).

Not incidentally, the PEF was also dependent on the State financially, although this contradicts the opening statement. The PEF was indeed in constant need for money as early as 1873. Appeals to funds were made on the religious importance that the Survey represented and were often carried out by senior personalities, as in the case of Prince Arthur's appeal at the occasion of the meeting of the Dover branch of the PEF in March 1872. The PEF would renew fund-raising appeals regularly, especially in the advertising columns of *The Times*. In an article printed in the latter on 7 September, 1880, it was reckoned that the survey, spanning between 1872 and 1877, as well as the printing of reports had cost £ 20,000. Even if the article mentions critical times for the PEF, which was often close to bankruptcy during the survey period, it never refers to State aid. And yet the State did help the Fund financially, partly as a consequence of the Safed attack.[10] The attack on the expedition party in July 1875 called the exploration to a stop, which endangered the completion of the Survey as the Fund was dangerously running short of money. In this context, repeated pleas were made by the PEF to the permanent under-secretary for Foreign Affairs, Lord Tenterden, asking the FO, on behalf of England, to do its duty and take action. Grove, honorary secretary of the Fund, wrote to Tenterden to that effect on 19 August, 1875:

> Our operations have all been carried out with money raised from private sources with great exertions on our part. The survey is most anxiously awaited by a large number of the religious people in England and their disappointment and mortification will be extreme if our progress is stopped when we are so near our termination. (PEF/DA/WS/CON/168)

Actually, even if interest in the survey amongst the general public had not died out, it was very clear that what the PEF needed most then, beyond a fair trial hopefully secured for by the Foreign Office, was money. This, the War Office understood. In his *Measuring Jerusalem: the Palestine Exploration Fund and British Interests in the Holy Land*, John Moscrop concedes that the War Office "could not give money directly to help the Fund" (2000, 114). Instead, he points out that the WO assisted the PEF in "providing in kind" help to draft the map (114), giving the example of the WO making available premises where the map could be produced, but also of seconding draughtsmen in June 1877 (119), after the survey had resumed. More surprisingly perhaps, the wages of the explorers who were Royal Engineers were paid for by the WO.[11] This is made clear in a letter dated 2 August, 1877 by Charles Wilson, who was now working for the intelligence branch of the War Office, to Walter Besant, the new PEF honorary secretary, that additional pay for the party men was purely a matter for the WO and did not have to be dealt with by the PEF.

At the annual meeting in 1875, the Archbishop of York remarked that the survey of the land of Western Palestine had somehow done away with its original aim, with "the poetry and romance of exploration," and as a result of many incidents—Safed being the most illustrious one (Jacobson & Cobbing, 168)—had been turned into "a most prosaic thing" (*PEF QS*, 1875, 113). For him, the perversion of the spirit of the survey came from the paradox that although the State, and the War Office in particular, refused to overtly finance the survey, it confidentially gave directions as to the manner in which the survey should be completed and, at times, relieved the PEF from its dire financial situation. Clearly, the Archbishop, and, more generally speaking, the whole PEF committee, resented an increasing dependence on the War Office, especially from 1876. In June 1877, in a letter to Besant, Lieutenant Conder, who had precisely been dispatched by the War Office, even expressed his astonishment at the sudden increasing interest in the Survey displayed by the War Office. He wrote that "[he did] not quite understand what led the WO to take so lively an interest in [their] proceedings but [he was] glad of it" (PEF/DA/WS/CON/304). This confidence invites us to consider the War Office's interest in the Western Survey as both geostrategic and timely.

The WO's vested timely interest in the PEF Western Survey

Edward Fox reviewing Moscrop's study wrote about the subordination relationship between the PEF and the WO in those terms:

> When it suited the War Office, the PEF was used literally as a front organization under which militarily important mapping and surveying expeditions were conducted, especially after the construction of the Suez Canal (1).

If it is clear that the relationship between the PEF and the WO could involve tension and even subordination, Fox is actually suggesting that the WO used the PEF expedition as a cover for military intelligence-gathering by Royal Engineers. Fox's statement is overall true, even if it must be kept in mind that the PEF never intended a military application for their map. To fully understand the cover artifice, one must remember that the WO had possessed its own Topographical Department since February 1855. It could thus come as a surprise to us that the WO did not undertake its own survey, especially if Palestine was a key strategic zone. Beyond the pragmatism which would forbid that a State survey be conducted in parallel with a private one, it is clear that a private survey would be less suspicious in the eyes of the Turkish Sultan, who was to agree to the expedition through a firman issued in 1872.[12]

Thus, the WO provided the competent staff with the required topographical and trigonometric knowledge to the PEF in exchange for keeping the original zinc plates of the map and "giving transfers of the zinc plates [to the Fund] on payment of the cost of the zinc" (PEF/MINS/ 12 March 1878). The relationship between the PEF and the WO was thus largely a gentlemen's agreement that the PEF had to accept because of its low finances.

In his article "Sacred but Not Surveyed" (2002), Haim Goren demonstrates that numerous surveys of Palestine had been attempted before 1871 and had failed, because contrary to the PEF Western survey, they had lacked support from an intelligence body.[13] With the reopening of the Eastern Question in 1875—by which the Concert of Europe had decided to prolong the life of the decadent Ottoman empire—the strategic value of the survey could not fail to appear to the War Office. The Safed incident in 1875, together with increasing tensions in the Ottoman empire—first due in July 1875, to revolts in Bosnia and Herzegovina and, in July 1876, to atrocities committed against Christians in Ottoman Bulgaria—proved to be incentives for the State, which now feared that the completion of the Survey was under threat. It became all the more urgent to get a complete map as there was no available one of the area which was detailed enough.

More important still, the map could be of imminent use to the WO should Britain—as a signatory of the 1856 Treaty of Paris which, after the Crimean War, guaranteed the integrity of the Ottoman empire, particularly against Russian interference—be led into war as a consequence of the Russo-Turkish War which had begun in 1877. To prevent an increased Russian influence at Constantinople, it was suggested that Britain could impose a protectorate over Palestine—and that too, would require a minute knowledge of the land that only the survey map could provide. Thus, by the time of the Safed incident, it was very clear from the War Office's point of view that this map was of high strategic value and that it should be completed as quickly as possible and at any cost, in spite of local crises and in spite of increased reluctance on the part of the Sultan to allow surveying.

It is also essential to bear in mind that the WO was conscious of the precious value of the data that were regularly being transferred to the Ordnance Survey in Southampton and was thus taking active steps to protect them, possibly even to the detriment of the PEF. The WO was indeed asking the PEF to delay the time when the map would be disclosed to the public (Moscrop 2000, 124) and so, to postpone the publication of its map until Eastern crises had calmed down. To secure its copyright over the map, the PEF had the WO promise that it would "not use [the map] except in condition of war" (PEF/ MINS/ 12/3/1878). Britain eventually did not take part in the Russo-Turkish War and the map was thus kept by the WO until the 1917 Palestine Campaign.[14] But by then, of course, the map had been reedited many times by the PEF and was thus not shrouded in secrecy anymore.

The political, military and geostrategic potential of the survey was understood by others in England than the WO. As early as 1876 indeed,[15] a small lobby close to political and military circles advocated the use of the map to delineate or simply extend the route of the Euphrates Valley Railway. The projected extension of the long-talked-about Euphrates Railway would involve the construction of a railway line between the Ottoman city of Mardin and Antioch in Syria, in the hope of connecting Antioch to Ismailia in Egypt. An extension of the Asia Minor and Euphrates Railway to Egypt would cross Western Palestine, surveyed by the PEF, and was thus conceived of as an extra argument to get funds to complete the survey. This idea was regularly circulated between 1876 and 1878 in local and national newspapers. The Euphrates Railway, sometimes considered as "the greatest highway"[16] and "the shortest route"[17] to British India, had first been evoked in 1867. No secure and definitive route had ever been agreed upon when the issue was raised again at the time of the Berlin Conference with the new route Baron Julius Reuter sent to the Premier, Benjamin Disraeli, in August 1878 (Hughenden Papers, Dep. Hughenden 75/3). The contemporary route was then slightly North of the extreme Northern zone surveyed by the PEF. As the political situation was particularly unstable in the area and as Russia was on the verge of controlling the Euphrates Valley—and why not the Jordan valley, which was on the Eastern border of the map—the Western Survey map was thought of as intelligence for the protection of the Palestinian access to the Suez Canal.[18]

Logically, the Western Survey required Biblical excavation and mapping. Nonetheless, mapping, as we have seen, was readily understood by the War Office as intelligence gathering. This created tensions between the PEF and the WO, in the same way that the awareness the Sultan had of the geostrategic potential of the Western Survey map created an even tenser international atmosphere, especially as the time of publication of the map coincided with the apex of this Eastern crisis. Once the Russo-Turkish War was over, the interest of the WO in the PEF enterprise dwindled, so much so that the Intelligence Department decreased its funding and did not help the PEF publish the Survey's results (Moscrop 2000, 124-5). Still, the political and military application of the map could not eschew its inherent objective, that is Biblical geography. Simply, Biblical geography applied to Palestine was particularly informed by two Biblical passages, respectively: *Jeremiah* 32: 26-44, in which the restoration of the Jews was foretold; and *Revelation* 16: 14-16, especially as the location of Armageddon where the last battle was to take place, was increasingly being identified as Megiddo.[19] Megiddo, situated in the plain of Esdraelon, was part of the Western Survey. The PEF enterprise, undertaken at a time when the Ottoman empire was shaken by numerous revolts, was thus read by prophesiers as a sign that the restoration was impending. Although restoration schemes were relatively "liminal" (Bar-Yosef 2002, 25), they became more visible after the publication of *Daniel Deronda* in 1876 by the mainstream novelist George Eliot and could more easily and more readily be read in contemporary events happening in the Ottoman empire.

A Biblical geographical understanding of the map for the colonisation of Palestine and the return of the Jews

Although the survey was not supposed to lead to prophecy, a number of comments were passed on the potential fruitfulness of the soil of the surveyed land, especially as the survey included key prophetical locations. In Protestant eschatology especially, the idea that the land of Palestine was to be made fruitful, especially by its colonisation, was understood as a step towards the restoration of the Jews in Israel, which would precede the second coming of Christ.

The PEF had never been either sectarian or wholly Protestant (the Jewish leader Sir Moses Montefiore was a member). This tolerance facilitated the appropriation of the survey map by various religious movements. Interestingly enough, three Royal Engineers who had been involved with the PEF—namely Warren, Wilson and Conder—expressed views about the future of Palestine. Charles Warren, who surveyed and excavated in Jerusalem (1867-1870), penned *The Land of Promise, or Turkey's Guarantee*, published in 1875, in which he recommended the colonisation of Palestine.[20] Warren advocated the restoration of "pure" Jews to Palestine under the supervision of a Great Power, suggesting implicitly that it was Britain's role. In *Tent Work in Palestine* (1879), Lieutenant Conder went even further, positing that the restoration of Palestine to "its former [Biblical] condition" (329-330) would require its occupation by some strong European power (332).

However, if the Western Survey map revived debates about the future of Palestine and its (re-)colonisation by Jews, it also served as a reminder for some Protestant eschatologists that Jews needed to be converted to Christianity before restoration. Otherwise, the Second Coming could not take place. It was the viewpoint adopted by the London Society for Promoting Christianity amongst the Jews, which had for president none other than the Earl of Shaftesbury, who had been president of the PEF. Upon becoming president of the PEF in 1865, Lord Shaftesbury declared:

> "Let us not delay to send out the best agents (...) to search the length and breadth of Palestine, to survey the land, and if possible to go over every corner of it, drain it, measure it, and, if you will, prepare it for the return of its ancient possessors, for I believe that the time cannot be far off before that great event will come to pass ..." (Quoted by Derek White)

Shaftesbury, who was speaking in his own name, but was also delineating a programme for the newly created Fund, was then encouraging the surveying of Jerusalem and, more generally, of Palestine, because it was preparatory work for restorationist colonisation schemes.

Shaftesbury's association between the timely survey and the closeness of revelation was taken up by British Israelites, who argued that Britain had Israelitish origins[21] and that Britons would return to Palestine together with Jews.[22] British Israelites had no connection whatsoever with the Palestine Exploration Fund, but reacted to their activities. Although controversial and marginal, British Israelism gathered momentum after the Eastern Question was reopened in 1875 and identified this Eastern crisis[23] as yet another sign of the end. As early as February 1877, the progress of the Western Survey was the object of an article entitled "Palestine Exploration in its Bearing on Israel's Restoration" in the recently founded *The Banner of Israel*. Other articles were published in *The Banner of Israel* about the PEF survey, which tried to demonstrate that it was a preparatory stage before the return of the Lost Ten Tribes of Israel, i.e. Britain, and the two tribes of Judah, i.e. Jews, to Palestine, and more particularly to Galilee.[24]

The Jewish Chronicle also viewed the Palestine expedition with great eagerness as it also linked this enterprise to the future return of Jews to Palestine, especially in the context of their repeated persecutions throughout Europe, and even more particularly in Russia. In a series of articles reprinted in the *PEF Quarterly Statement* in April 1880, there was a letter by Conder addressed to the editor of the *Jewish Chronicle* (*PEF QS*, 1880, 114-118). Enlightened by his experience as surveyor of Western Palestine, Conder valued Laurence Oliphant's colonisation scheme in eastern Palestine, and more specifically in the land of Gilead, but thought it could only become successful if the railway extended to this region.[25]

Oliphant imagined a colonisation scheme for Palestine that would involve the return of Jews and for which he had received Lord Beaconsfield's support.[26] Interestingly enough, he pointed out that this scheme could profitably be backed up by Protestants who would see this as a fulfilment of prophecy, although he wanted to distance himself from popular religious theory.[27] Displaying what Eitan Bar-Yosef named the "uneasy relationship between the millenarian and the imperial" (2002, 27), Oliphant clearly played with the restorationist interpretation his scheme could be endowed with, as he thought he could get more support if he presented his plan as not being solely political, or worse strategical. In a biography entitled *Memoir of the Life of Laurence Oliphant and of Alice Oliphant, His Wife* (1891), the author quotes a letter by Oliphant dating back to 10 December, 1878, in which he explained the genesis of his scheme:

> My Eastern project is as follows: To obtain a concession from the Turkish government in the northern and more fertile half of Palestine, which the recent survey of the Palestine Exploration Fund proves to be capable of immense development. (285)

Oliphant always showed keen interest in the Western Survey, whose results comforted him in the choice of the location for his colonisation scheme. Oliphant was also in contact with Conder, who regularly published updates on the latter's scheme in *The Jewish Chronicle*.[28]

To conclude, despite the seemingly scientific and purely geographical name "Western Survey," the PEF expedition was thought of as being multi-purpose, be it by the Palestine Exploration Fund itself, or by other organisations, especially the War Office and the Foreign Office. In "Sacred but Not Surveyed," Haim Goren sees the map as being first of geographical relevance. He implicitly suggests that the "georeligious"—that is the restorationist and the eschatological— and the "geostrategic"— meaning the political and the imperial—converged in and enriched the PEF survey because of the conception of geography as an evolving nineteenth-century science. This multiplicity of possible readings of the Western Survey map, as well as of potential publics, is visually present on the map itself through the mention of different types of names for each site (Old Testament, New Testament, Apocrypha, Josephus, Talmudic and modern Arabic names[29]). Of course, the different understandings of the map could come into conflict, herewith revealing unofficial agendas and power struggles—for instance, between the WO and the PEF, or between the PEF and British Israelism. However, more often, as displayed by the schemes and opinions of pivotal figures such as Charles Warren, Claude R. Conder or Laurence Oliphant, the "georeligious" and the "geostrategic" partook of one and the same conception of the future of Palestine and were actually complementary.[30] *In fine*, what gathered together these different readings of "the perfect map" was the hope it inspired and the certainties accurate topography provided. Undertaken and disclosed at a time of dire crisis in the Eastern Question, the survey could then be reinvested by the various circles who took interest in the Western Survey—be they scientific (the British Association), political and imperial (the State), or eschatological (Protestant and Jewish)—with their own hopes for times ahead.

Works cited

Primary sources

Archives
Hughenden Papers, Bodleian, Oxford.
Palestine Exploration Fund Archives, PEF, London.
War Office Papers, National Archives, Kew, London.
The Hansard.

Newspapers & periodicals
Glasgow Herald
Northern Echo
Palestine Exploration Fund Quarterly Statement
The Aberdeen Journal
The Banner of Israel
The Jewish Chronicle

The Times

Other works
Map of Western Palestine, in 26 Sheets, from Surveys Conducted for the Committee of the Palestine Exploration Fund, by Lieuts. C.R. Conder and H.H. Kitchener, during the Years 1872-1877. London: [s.n.]. 1880.
Cazalet, Edward. 1878. *The Eastern Question: An Address to Working Men, with Map Showing the Projected Line of the Euphrates Valley Railway.* London: Edward Stanford.
Conder, Claude Reigner. 1879. *Tent Work in Palestine: A Record of Discovery and Adventure*, Vol. 2. London: Richard Bentley & Son.
Oliphant, Laurence. 1880. *The Land of Gilead: With Excursions in the Lebanon.* London & Edinburgh: W. Blacwood & Sons.
Palestine Exploration Fund, Committee of. 1873. *Our Work in Palestine Being an Account of the Different Expeditions Sent out to the Holy Land (Since the Establishment of the Fund in 1865).* London: Bentley & Son.
Patrick, Andrew William. 1857. *The Euphrates Railway: the Shortest Route to India.* London: Effingham Wilson.
Wale, Burlington B. 1893. *The Day of Preparation; Or the Gathering of the Hosts to Armageddon: A Book for the Times.* London: E. Stock.
Warren, Charles. 1875. *The Land of Promise, or Turkey's Guarantee.* London: George Bell & Sons.
Wilson, Margaret Oliphant. 1976. *Memoir of the Life of Laurence Oliphant and of Alice Oliphant, His Wife.* New York: Ayer Publishing.

Secondary sources

Bar-Yosef, Eitan. 2002. Christian Zionism and Victorian Culture in Goren, Haim. Sacred but Not Surveyed. In *Imago Mundi*. London: [s.n.], Vol. 54: 87-110.
Hopkins, I.W.J. 1968. Nineteenth-century maps of Palestine: dual-purpose historical evidence. In *Imago Mundi*. London: [s.n.], Vol. 22: 30-36.
Jacobson, David & Felicity Cobbing. 2005. 'A Record of Discovery and Adventure': Claude Reign Conder's Contributions to the Exploration of Palestine. In *Near Eastern Archeology*. Boston: Boston University. Vol. 68. Issue 4: 166-179.
Kobler, Franz. 1956. *The Vision Was There: A History of the British Movement for the Restoration of the Jews to Palestine.* London.
Online at: < http://www.britam.org/vision/koblerpart4.html> accessed 11/01/08

Lippman, V.D. 1988. The Origins of the Palestine Exploration Fund. In *Palestine Exploration Quarterly*. London: Maney. Vol. 120. Jan.-June: 45-54.

Moscrop, John James. 2000. *Measuring Jerusalem: The Palestine Exploration Fund and British Interests in the Holy Land*. London & New York: Leicester University Press.

White, Derek. Christian Zionism. Onlineat: <http://www.zionism-israel.com/christian_zionism/Christian_Zionism_history.html> accessed 11/01/08

CARTOGRAPHY IN THE AGE OF SCIENCE: THE MAPPING OF IRELAND IN THE NINETEENTH CENTURY

VALÉRIE MORISSON

The scientific outlook that prevailed all through the 19th century affected several disciplines including cartography. As projective geometry developed, maps relinquished their decorative function and their kinship with engraving or painting to become more informative. They reflected a new drive towards standardisation and codification. Alongside the geographical map, geological maps spread and exemplified a newly felt need for comprehensiveness and exhaustiveness:

> Maps made the new geography the model for a scientific revolution that replaced ancient authority with experience and mathematical description. (...) Maps were the undeniable makers and markers of modernity, the signs, as well as the tools of a distinctly new age. (Helgerson 2001, 241)

The mapping of Ireland, undertaken by the Ordnance Survey (O.S.) as early as 1820, testifies to the latest developments in mathematics and geometry at that time. The O.S. also illustrates the modernization discourse that underlay British imperialism. As a matter of fact, it was part of a larger trend at a time when demographic surveys and the collection of numerical data stemmed from the need for reform. The systematic mapping of colonized regions and the elaboration of cadastral maps fell within the province of colonial expansionism, then of imperialism. New geographical institutions and scientific investigations granted colonial ventures more legitimacy. Indeed, the accuracy of the map accounts for its power, so much so that scientific knowledge and political power are interwoven. If maps were designed and used by the ruling classes and remained strangers to popular culture, it is because they were levers of power and not instruments of resistance. However, in the case of the O.S, a purely colonial reading has to be reconsidered since the Anglicization of the landscape and the modernization that was to ensue did not exclude a renewed interest in things indigenous.

Map-making under scrutiny

From the 16th century onwards, map making has obeyed scientific rules. To map a territory, you must look at the land, understand it, and organize the data collected through perception. This requires the use of systematic surveys, transcriptions, and scientific projections. A map is the outcome of a rigorous classification: a faithful depiction of space relies upon a scientific turn of mind capable of organizing and structuring this space.

Several contemporary geographers, working in the wake of Roland Barthes, Michel Foucault and deconstructivist thinkers also consider the map as a text, with an author, a receptor, a code and a message. Mark Monmonier complains that we consider maps, like figures, as sacred images while they seldom deserve an unrestrained respect (Monmonier 1991, 27). Mapping inevitably entails distortions: the choice of scale and symbols, the type of projection, or the lines may alter our perception of space. In order to represent a place, the cartographer has to simplify, reduce or enlarge, modify the scale or ignore some details. Therefore, a map, however accurate, is nothing but one given representation among many of a given space, a subjective representation.

A reader of Roland Barthes, Harley holds that cartography has never been an autonomous or watertight science, nor is it impervious to the political machinery of knowledge (Harley 1992, 231-247). Maps belong to a cultural system: they are shaped not only by the scientific rules of geometry but also by norms and values bred by social traditions (Harley 1992, 64). Harley's analysis of the map finds its origin in Michel Foucault's assumption that knowledge means power. Instead of seeing the map as the result of measurement and topographical surveys, Harley sees it as a rhetorical device. Far from reflecting reality, the map therefore generates a new reality (Harley 1992, 85). A map is an appraisal, an assessment, an instrument of persuasion. Mark Monmonier notes that "most symbols and geographical projections can only serve very specific communication goals. (…) The map is what it is because its author knows in advance what it should be like." (Monmonier 1991, 12, 76). The map is indisputably induced by authorial decisions, whether it be the selection, classification, generalization of data or the choice of symbols: "The authors of a map are entirely free to choose various details, measures, coverage area or symbols that will serve best their interest or reveal most efficiently their conscious or unconscious bias" (Monmonier 1991, 25). Signaling either a castle or a church by a large and visible sign is in no way ideologically neutral.

Topographical surveys, measurements thanks to triangulation, rigorous mathematical projections have never been able to delete the political symbolism embedded within the map. The knowledge spread by the map is not merely topographical, even though maps are conceived first and foremost as scientific systems. Maps admittedly respond to the precise needs of a political regime and become intellectual weapons geared towards the acquisition of power and its establishment (Monmonier 1991, 25). Some pertain to what Foucault calls surveillance acts. What is more, maps enable colonial kingdoms or states to assert their dominion over their colonies:

> Maps have made it possible for Europe to easily dismember Africa and other wild lands, to impose its rights upon the countries and their ressources and to ignore pre-existing political and social structures. (Monmonier 1991, 137)

Much has been written about the scientific discourse backing colonial ventures in non-white countries. However, Ireland seldom comes under scrutiny when the history of British imperialism is at stake.

Mapping British Ireland: the outlines of a colony

While the position of Ireland in the orbit of England was firmly established—by Ptoleme and Mercator (1570)—the island itself was only vaguely represented at the time. As early as the late 1540s, maps were part and parcel of the administrative machinery of modern England. As far back as the first colonisation of Ireland, the mapping of the country altered the organisation of space. From 1540 to 1750, the island was mapped and depicted on several instances by the new elites—landowners, governors, engineers, cartographers and other officials. As Bernhard Klein noted, during the Elizabethan period, the first English maps showed either the unity between Ireland and England or enhanced their differences. On the very first English map, Ireland is represented as an irregular oval shape containing many blank spaces vaguely featuring stylized mountains. The 1520 *Cotton Map* testifies to the control of the Eastern coast and foreshadows the impending colonisation of the West.[31] Places under English control and the lands of allied squires are clearly indicated. In the middle of the 1560s, a general map of Ireland by John Goghe confirmed that the East and the South were fairly well-known while the west and the north remained largely unknown. The map of Ireland printed in Abraham Ortelius' *Teatrum Orbis Terrarrum* (1573) was still very inaccurate. However, the progress of England in the sciences of cartography and navigation led to a more accurate representation of Ireland. In the late 16th century, English cartographer Robert Lythe undertook the mapping of the south of the island. His map enabled the viewer to identify the castles, forts, churches and the main families' domains. Lythe also contributed to delineating county boundaries precisely. In the opening years of the 17th century, English war artist and cartographer Richard Barlett drew a minute military map of Ulster that featured placenames. In 1685, William Petty provided a precise view of Ireland. His *Hiberniae Delineato* included 10,000 English names and depicted an efficiently governed, carefully colonized island.

As Gerry Smyth makes clear, all through the 18th century, colonisation and science went alongside: the map was an illustration and an assertion of the colonial power (Smyth 2001, 49). The blanks on the map hindered decision-making and were therefore a hurdle in the running of the colonies. Land surveyors, soldiers and administrators paced the territories:

> Space is primarily what is known, described, walked through and mastered. It must be the setting of explorations and victories. Power is then asserted thanks to its spatial occupation as it becomes a territory. Power will therefore always endeavour to know its space better, to discover new space likely to widen its basis and to deepen its grounding. (Regnauld 1998, 86)

Despite their increasing scientific accuracy, maps still showed deficiencies. However, in England, the enclosures had paved the way for improvements in cadastral maps. H. S. Homers' 1766 *Essay on the Nature and Method of Ascertaining the Specifik Shares of Proprietors upon the Inclosure of Common Fields* finalized a method presiding over the drawing up of cadastral maps. Later on, the 1836 Tithe Commutation Act listed some legal requirements in terms of precision and accuracy for a map to be officially registered. Cadastral maps were a prerequisite to the valuation of lands and permitted more efficient running of the lands as well as greater trade opportunities (Kain & Baigent 1984, 8). By clearly delineating the boundaries of domains and estates, cadastral maps affected the relation between landowners, tenants and peasants: "By teaching landowners the precise extent and nature of their holdings, maps pushed Europe's agrarian economy away from the feudal manor and toward the capitalist market" (Helgerson 2001, 241). In 16th century

England, large-scale cadastral maps encouraged rural capitalism. They led to a more profitable exploitation of the land, higher rents, enclosures, drainage and filling up. The surveyors and the landowners, working hand in hand, encouraged a capitalist conception of agriculture (Harley 1992, 30-31). Topographical surveys enabled the authorities to draw up a legally constraining image of space. As geographical conventions were known only to the elites, peasants could hardly question the delineations that illustrated the laws operating a new distribution of the territory.

The English implementation of a system of exploitation that foreshadowed capitalism, as well as the increasing centralization, modified the Irish landscape. In order to ground these modifications into a new colonial culture, the English set up narratives and representations, such as the map, apt to convey it. Maps were first used in reconnaissance missions and provided military information. They were subsequently used to pacifiy, civilize and exploit newly-conquered colonies. Maps therefore asserted the conquest of new territories and extolled the virtues of the empire: "All along the exploration age, European maps gave a one-sided image of ethnic conflicts and supported Europe's divine right to appropriate new territories" (Harley 1992, 38-39). In 1580, after the seizure of Irish rebel Gerald Fitzgerald's lands and the plantation of Munster, the land was mathematically divided regardless of natural geographical features. The estates were registered and the measures standardized. The acre replaced the medieval Irish units though in most cases the colonial power dared not ignore former local ones. The map indisputably became an instrument of observation and surveillance reflecting Anglo-Irish relationships:

> In 17th century Ireland, for example, the simple fact that the surveyors who worked for English landowners failed to record the natives' cabins on their otherwise precise maps has nothing to do with the scale of this sort of dwelling but can be accounted for by the religious and social tensions rife in the Irish countryside. (Harley 1992, 37)

A series of oppositions underlie the drawing of a map (centre vs. periphery; latin vs. non-latin; modern vs. backward-looking; cilivized vs. primitive; urban vs. rural, etc.). The so-called backwardness of the Irish was reflected onto the English map. Very much unlike the neat and precise landscape engraved by the English, the Irish landscape was inevitably conceived as disorderly and shapeless. Consequently, the maps sharpened the perception that the English culture was superior to the Irish one. The Irish knowledge of the land was transmitted through traditions and folklore (tales, poems or ballads) and linked to genealogy. While the

English outlook on the island was a centralized one, the geographical knowledge of Irish people was regional and local. Ireland was figured out as a collection of various cultures. The Irish lords had their own experts or *criochairi* who would define boundaries revealing a clan-shaped perception of space as enshrined in the Brehons. The maps established by the barons often were the outcome of a compromise between the settlers and the local occupants:

> Much of the time, the systematic programme of confiscation and resettlement—as for example during the wars and plantations of the sixteenth and seventeenth centuries—was superimposed on and somehow reconciled with indigenous spatial knowledge, thus securing the 'double benefit' of co-opting the native past and diminishing the likelihood of resistance to the new order. (Smyth 2001, 49)

For all that, the authorities failed to register all the allotted plots and to provide absolutely reliable data so that taxation was reckoned mostly unfair. As a result, in the 1820s, the Irish landowners strove to convince Westminster that a thorough survey and valuation of the land was needed if taxation was to become fairer. The new maps would also guarantee greater political stability.

The double-edged mission of the Ordnance Survey

In 1824, in London, it was eventually decided that more six-inch maps of Ireland would be drawn in order to revise valuation. The O.S.'s mapping of the country is a pivotal event in Irish history. The mission was led by Royal engineer Thomas Frederick Colby. Richard Griffiths, a widely recognized geologist, oversaw the launch of the operations and contributed maps. Never before had such a large-scale scientific survey been undertaken in Ireland. Up to 2139 men worked on the project including officers, soldiers and local workers, draughtsmen and collectors. The project was twofold. On the one hand, English engineers were asked to draw maps and to standardize place names. On the other hand, memoirs were initially to complement the maps in an attempt to testify to the richness of the Irish culture in a scientific way.

Underlying the new scientific undertaking was the desire to circumscribe the otherness of Ireland in a more thorough way. Gillian M. Doherty, who provided a most thorough analysis of the O.S.'s project, states that while in the beginning of the modern era, maps were meant to either firmly establish and strengthen governments, elites and aristocrats or to conduct military operations, in the 19th century, the maps bore the influence of utilitarianism. Their main function was to control the population and to utilize the resources more efficiently (Doherty 2004, 12-14). To do so, the O.S. had to standardize local practices, especially when it came to measures. In the closing years of the 16th century, the English acres had already been imposed upon the population and had changed Irish people's habits. For all that, the mapping of Ireland was not a crudely colonial project but made it clear that Ireland and England were to be culturally and economically united. One should indeed keep in mind that the Union Act had been voted in 1801.

Firstly, the O.S. maps turned the townland, or *baile* in Irish, into the main territorial administrative division replacing older Irish sub-divisions. The administered space was reorganized: new townlands were created, others were significantly modified, enlarged or deleted. Secondly, the neat and clear outlines on the maps indicate estate boundaries as defined by their owners rather than by cartographers. Space is conceived in social terms as the maps enhance the equation between social and geographical divisions. The lands around the larger mansions are easily visible; more than 10,188 individual country houses are featured and named, some of them having no surrounding park. 2,596 estates of 50 acres or more are also materialized (Graham–Proudfoot). The maps supplemented the 1841 census. Thirdly, the maps no longer have a decorative function and become scientific landprints. Most maps are minute but abstract depictions in black and white; there is no attempt at suggesting the real landscape through colours or gradations. Lastly, the names of the cartographers and contributors, as well as the indication of their hierarchical position, are given on each print, which bespeaks the eagerness to authorize and legitimize the maps.

Historians give diverging accounts of the English undertaking. The O.S. gathered English and Irish experts and workers but very few of the landowners whose interests were served by the project were Catholics. American historian Mary Hamer stresses the domination of Anglo-Irish landowners to the detriment of Catholics who were thereby kept away from the project (Hamer (a) 23-24). Hamer states that the maps were capitalist instruments strengthening the control of the means of production and English colonial policies in Ireland (Hamer (a) 23): "an official Ireland was produced, an English-speaking one, with its own ideology of Irish space" (Hamer (b) 185). Ireland, as depicted in those maps is

> both familiar and foreign, a comforting image of a recognizable English spatiality and an object of ethnographic otherness. [...] The Ordnance Survey etches a placid, idyllic topography whose visual specificity creates a convincing portrait of an ordered and peaceful nation under the guiding

hand of a benevolent colonial administration. In particular, the map's emphasis on Ascendancy manor houses projects a kind of unionist geography. (Hegglund 2003, 173)

The choice of scale enabled the state to turn what was still a wild land into a civilized territory thriving thanks to colonial administration. The names of newly settled colonizers featured on the maps seem to confirm Hegglund's conclusion. However, the maps recorded some indigenous features too, for instance the traditional *clachan*, derived from ancient circular forts. As a consequence, the map can also be construed as an attempt to enhance local architecture. The double-edged nature of the O.S. and its political ambiguity are even more blatant when one considers the memoirs.

As he supervised the beginnings of the operations conducted by the O.S., Captain James Pollock, Colby's assistant, advised his agents to "pay attention to the social condition and habits of the people, to antiquities and traditionary recollections of all kinds or to natural history in all its branches" (Doherty 2004, 17). His successor, Captain Thomas Larcom, equally versed in history and culture determinedly encouraged the publication of memoirs that would supplement the information provided by the maps. The collections of the O.S. Library at the time testify to the scholarship of its members. Anglo-Irish scholars eager to safeguard Irish culture also took part in the memoir project. The 19th century's craving for statistics and systematic collection of data induced a preference for first hand testimonies and information that guaranteed the scientific quality of the studies. Consequently, contributions and reports by Irish-speaking collectors were particularly sought-after. The memoirs were nonetheless short-lived: in 1840, they came to an abrupt end after a controversy broke out. Irish scholars and Irish newspapers backed Larcom's project but proved unable to curb Peel's decision to restrain the mission of the O.S. to cartography. However, if Thomas Larcom failed to produce memoirs that would have built up a more accurate knowledge of Irish traditions, some cultural elements transpired on the maps, notably through the printing of placenames.

Considering the project in its entirety, one is therefore entitled to question the English government's earnest interest in Irish vernacular culture. Gillian O'Doherty views the O.S. project as a scientific and scholarly undertaking likely to save an already disappearing culture. Back in the 1830s, archeology was still tentative and spotty. Irish people paid less attention to their heritage and neglected the preservation of their monuments. The Irish language itself was seriously threatened. John O'Donovan, member of the O.S., feared local customs might soon disappear. The O.S. and the Royal Irish Academy worked hand in hand to make old manuscripts and ancient heritage better known to large audiences. George Petrie, one of the most renowned antiquarians promoted the O. S.'s research on the prehistoric site of Tara in two lectures that he delivered at the Royal Irish Academy in 1837. The memoirs contributed to the development of archeology, spelling, genealogy and philology, that is to say sciences that would fuel the forging of a strong national identity during the Celtic Revival. As a matter of fact, the late 19th century nationalists reached into the data collected by colonial servants so that the O.S. ethnological department can partly be held responsible for the awakening of national awareness among the Young Irelanders for example (Bulson 2001-2002, 92).

Colonialism indisputably spurred such enquiries into local customs. If Larcom and Colby were intent on safeguarding Irish culture and were sensitive to poverty in Ireland, they were not impervious to the prevailing colonial discourse. As a matter of fact, if they did not buy the Malthusian theory—that poverty in Ireland was due to over-population—and considered that a bad administration of the island was responsible for Ireland's backwardness, they saw colonisation as a means of bringing progress and prosperity to the island. Some of the contributors to the memoirs claimed that poverty could be accounted for by indigenous if not racial innate features. Most scientific surveys had a racial bias. Scientists strove to establish a connection between ethnicity and economic performance. The collection of data aimed at assessing the position of Ireland on a development scale that summed up social evolutionism. Most scientists held that countries evolved at various rhythms on a universal, supposedly natural time line. Ireland was scientifically labeled as "primitive" and henceforth positionned below England.

One is therefore tempted to follow Mary Hamer's statement that ethnology is based on a now embarrassing dichotomy between the scholarly observer and the objectified, uncivilized other:

> Everything about Irish life that was 'othered' by process of meticulous record was liable to define itself as primitive, if not savage, degenerate: mutely pleading, indeed, for the imperial helping hand of civilized England. (Hamer (b) 187)

The Irish scholars who were associated with the project but shunned away from any executive power were a means of softening the imperial policy. The conclusion that Hamer's reader is bound to draw is that the collection of anthropological and ethnological data constituted a strategical compensation for the actual redefinition of the territory to the benefit of the English. Hamer's final contention is that the O.S. map was both a means of enhancing and alienating the past (Hamer (b) 187).

The O.S. was admittedly a pivotal articulation between modernity and tradition:

> The map/names produced by the O.S. were textual evidence of the survival of native knowledge within a new political dispensation—the Union—and thus a fitting symbol of the way old and new, Irish and British, could be meaningfully reconciled. (Smyth 2001, 51)

Eric Bulson underscores the paradoxical nature of a project that records folklore and heritage while heightening the pace of their demise.

Placenaming and the preservation of Irish identity

Of particular interest in this respect is the issue of place-naming: the O.S. recorded Irish placenames while Anglicizing some of them. Thomas Colby considered place-naming in Ireland as unscientific. The Viking, Norman and Elizabethan cultures had all left an imprint on Irish placenames. They had changed with the successive waves of invasions without the changes being systematically registered. Unsurprisingly then, the fluctuating spelling of place names was looked upon by the English as a hurdle to proper management of the land. Their translation was considered all the more useful as the maps were made for the English:

> Because the overwhelming majority of those who would be using the maps were English-speaking, the Survey tolerated inexact renderings of Irish pronunciation in order to achieve a written form that would not look too outlandish to English readers. (Hamer (b) 194)

Some data-gatherers were recruited to collect all available information as to placenames. Linguists, anthropologists, as well as men of letters took part in this vast enterprise. John O'Donovan was in charge of recording Irish placenames before they disappeared. Landowners, royal servants and Irish-speaking laymen decided on the official name of Irish places. Both O'Donovan and Edward O'Reilley, an etymologist and a lexicographer, studied controversial cases. The whole project triggered a new interest in toponymy that may account for the publication of P. W. Joyce's *Origin and History of Irish Names of Places* in 1869. Toponymy is undeniably part and parcel of vernacular culture: "It seems, in fact, that no place may exist without the bestowing of a placename, or rather that the place comes to existence only once it is given a name" (Taverdet 2007, ii). Placenames reflect a constructed image of national or regional identity. Unsurprisingly enough, the correlation between toponymy and Irish colonial identity is at the core of Brian Friel's much acclaimed post-colonial play.

First staged during Derry's Field Days in 1980, *Translation*, Brian Friel's most famous play, unfolds in 1833 rural Ireland, in Ballybeg, Co. Donegal, during the first surveys. Though written in hindsight, *Translation* enables the reader to grasp what was at stake at the time of the O.S. and echoes the debates then going on over the anglicization of many placenames. Captain Lancey, an English cartographer—"a perfect colonial servant" (Act 1, Scene 1)—is seconded by Owen, in charge of the translations, and Lieutenant Yolland, a spelling expert. The play pits the mathematical and technical discourse of the English against the poetry of Irish lores. Lancey explains:

> His majesty's government has ordered the first ever comprehensive survey of this entire country—general triangulation which will embrace detailed hydrographic and topographic information and which will be executed to a scale of six inches to the English mile. [...] This enormous task has been embarked on so that the military authorities will be equipped with up-to-date and accurate information on every corner of this part of the Empire. (*Translation,* Act 1)

While for the English the landscape is to be measured and turned into properly ordered territory, for the Irish it is narrated and turned into culture. William J. Smyth notes that

> their concepts of territory and 'proper' territorial organization are compared with Irish ways of remembering, narrating, understanding and using landscape and territory to define, defend and use their patrimonies. (Smyth 2006, xx).

The creation of mountains, rivers and lakes is recounted in two Irish texts: the *Lebar Gabàla* and the *Dindshenchas* (Book of Placelore). In Irish popular culture, placenames are intertwined with family names:

> Named places, sometimes defined and identified by a natural feature (a mountain, a bog, a strand, a river, a natural well, etc.), did not generate simply local lore, but also a topography intimately bound up with families, ownership, genealogy... Places, placelore, placenames: the landscape of Ireland was seen and read by the Irish through powerful cultural lenses. (Foster 1997, 45)

The Gaelic *Dindshenchas* exemplify the profound connection between identity and landscape:

> The minute naming of Ireland's landscape features [...] is a function of a long-running narrative, fabricated, transformed, told and retold by generations of dwellers-cum-story-tellers who have occupied this land over the millennia. (Smyth 2006, 3).

As a consequence, the anglicization of Irish placenames was perceived as a blow dealt to Irish collective and individual identity. In Brian Friel's play translating often rhymes with distorting. As Yolland puts it, "something is being eroded" (*Translation* Act 2). For all that, other characters support the standardization of Irish placenames. Maire wishes she could speak proper English and reminds her fellow-villagers that Daniel O'Connel himself contended that Irish hampered progress. Hugh, the schoolteacher, eventually questions the preservation of the Irish language: "it can happen that a civilization can be imprisoned in a linguistic contour which no longer matches the landscape of... fact." (*Translation* Act 2). Brian Friel's text bears witness to the ambiguity of the O.S. and the questions it raised in the 1980s, at a time when Ireland was reconsidering its national narrative. The play, written 150 years after the first O.S. mapping of Ireland, provides the readers with no univocal interpretation of the historical situation. Though the maps and the names they carry are presented as colonial endeavours, the nationalist thrust epitomized by Manus is openly checked by other characters. Delving into the history of the O.S. maps of Ireland, one inevitably reaches the conclusion that a fair account of the events requires that we scrutinize the behaviours not of communities but of individuals. In the final analysis, we may consider the play as an invitation to revise the historical narrative, to challenge the nationalist assumptions, and to overcome the colonial divide.

The quarrels over the project opposed the advocates of Irish culture to the champions of modernization. However, this divide and the colonized/colonizer dichotomy did not strictly overlap. The political debates surrounding the memoirs evidence the lack of consensus among the O.S. staff. Besides, science permeated the field of cartography, and thereby became an instrument in the hands of the ruling classes, but equally percolated amongst the folklorists whose concern was to preserve Irish culture without negating the necessity of modernizing the economy of the island. One must indeed be aware that the discourse of modernization was not the prerogative of the English; modernization should not be equated solely with evolutionary progress and European imperialism. In a country not sure whether it has overcome colonialism and entered post-colonialism, the heated debates over the mapping of Ireland are not closed: the questions raised by the O.S. lead us at the core of the historiographical turmoil of the 1980s and 1990s. Hamer and Smyth strove to do away with the modernization theories resting on a simplified dichotomy between traditional (i.e. Irish) and modern (i.e. English). The reactions against the O.S. should in fact be reinterpreted within a larger framework, that is to say not as an opposition to the English state apparatus fuelling the nationalist movement but rather as the outcome of complex interactions between a traditional society and an imperialist world economy.

Works cited

Andrews, J. H. 1974. *History in the Ordnance Map: An Introduction for Irish Readers*. Dublin: Ordnance Survey.

—. 1975. *A Paper Landscape: The O. S. in Nineteenth-Century Ireland*. Oxford: Clarendon.

—. 1997. *Shapes of Ireland: Maps and Their Makers 1564-1839*. Dublin: Geography.

Boland, Eavan. 1994. That the Science of Cartography Is Limited. In *A Time of Violence*. New York: Norton, 7-8.

Bulson, Eric. 2001-2002. Joyce's Geodesy. *Journal of Modern Literature.* Vol. 25. Winter.

Rivière, J.L. and Llopès, M. 1980. *Cartes et figures de la terre*. Paris : Centre George Pompidou.

Doherty, Gillian M. 2004. *The Irish O.S., History, Culture and Memory*. Belfast: Four Courts Press.

Foster, John Wilson. 1997. Encountering Tradition. In *Nature in Ireland : a Scientific and Cultural History*. Dublin: Lilliput Press.

Foucault, Michel. 1972. *The Archaeology of Knowledge*. Trans. A. M. Sheridan Smilth. New York: Pantheon.

Friel, Brian. 1981. *Translation*. London: Faber and Faber.

Graham, B. J. and L. J. Proudfoot. 1993. *An Historical Geography of Ireland*. London, San Diego: Academic Press.

Hamer, Mary. 1989. Putting Ireland on the Map. *Textual Practice* 3.2.

—. The English Look of the Irish Map. *Circa*: Belfast. n°46: 23-24.

Harley, Brian. 1992. Deconstructing the Map. In *Writing Worlds: Discourse, Text and Metaphor in the Representation of Landscape*. Eds Trevor Barnes et James Duncan. London, New York: Routledge.

Hegglund, Jon. 2003. Ulysses and the Rhetoric of Cartography. *Twentieth Century Literature*, Summer.

Helgerson, Richard. 2001. The Folly of Maps and Modernity. In *Literature, Mapping, and the Politics of Space in Early Modern Britain*. Ed. Andrew Gordon and Bernhard Klein. Cambridge: Cambridge University Press.

Howes, Marjorie. 2000. Goodbye Ireland I'm Going to Gort: Geography, Scale, and Narrating the Nation. In *Semicolonial Joyce*. Ed. Attridge and Howes, University of York, Rutgers University, 58-77.

Kain, J. P. and E. Baigent. 1984. *The Cadastral Map in the Service of the State, a History of Property Mapping*. Chicago: The University of Chicago Press.

Klein, Bernhard. 1992. Partial Views: Shakespeare and the Map of Ireland. *Early Modern Literary Studies*. 4.2, September.

Le Squère, Roseline. 2006. Analyse des perceptions, usages et fonctions des toponymes actuels des territoires ruraux et urbains de Bretagne. In *Noms propres, dynamiques identitaires et socio-linguistiques*. Ed. Francis Manzano. Rennes: Presses Universitaires de Rennes.

Monmonier, Mark. 1991. *How to Lie with Maps*. Chicago: University of Chicago Press.

O'Brien, Eugene. 2002. *Examining Irish Nationalism in the Context of Literature, Culture and Religion: a Study of the Epistemological Structure of Irish Nationalism*. Dublin: Edwin Mellen Press.

O'Cadhla, Stiofain. 2007. *Civilizing Ireland: O. S. 1824-1842, Ethnography, Cartography, Translation*. Dublin: Irish Academic Press.

Regnauld, Henri. 1998. *L'Espace, une vue de l'esprit?* Rennes: Presses Universitaires de Rennes.

Richards, Thomas. 1993. *The Imperial Archive: Knowledge and the Fantasy of Empire*. London: Verso.

Roy, James Charles. 1997. Landscape and the Celtic Soul. *Eire-Ireland : an Interdisciplinary Journal of Irish Studies*. 31. 3&4. Fall, Winter.

Smith, A. 1998. *Landscapes of Power in Nineteenth Century Ireland. Archaeological Dialogues*. Vol. 5. n°1: 30-35.

Smyth, Gerry. 2001. *Space and the Irish Cultural Imagination*. UK: Palgrave Macmillan.

Smyth, William J. 2006. *Map-making, Landscape and Memory, a Geography of Colonial and Early Modern Ireland c.1530-1750*. Cork: Cork University Press, in association with Field Days.

Taverdet, Gérard. 2007. (foreword to) *Espace Représenté, Espace Dénommé, Géographie, Cartographie, Toponymie*. Valenciennes: Presses Universitaires de Valenciennes.

Whelan, Kevin. 1992. Beyond a Paper Landscape—John Andrews and Irish Historical Geography. In *Dublin City and County: From Prehistory to Present*. Dublin: Geographical Publications.

Wood, Denis. 1992. *The Power of Maps*. New York: Guilford.

PART II:

CLASSIFYING PLANT AND HUMAN SPECIES

WESTERN SCIENCE, COLONISATION AND REIFICATION: A CASE STUDY OF THE UTILITARIAN USE OF BOTANY IN DAVID MALOUF'S *REMEMBERING BABYLON*

CHRISTINE VANDAMME

The reason why botany has such pride of place in David Malouf's internationally acclaimed *Remembering Babylon* (1994) is really twofold: first the novel is about the early days of Australian settlement, and Australia's destiny and development are indistinguishable from a very influential political figure, Joseph Banks, who also happened to be a botanist; secondly, botany is a science which poses the question of naming and nomenclature with a particular acuteness and as such, it stages in miniature form wider scientific and political issues revolving around language, terminology and power.

The question of the classification of plants and the names they should be given is closely linked to the evolution of the discipline as Foucault has shown in his chapter on botany in *Les mots et les choses* (2002) or Jean-Marc Drouin in *L'herbier des philosophes* (2008). But as we will show, it is also and more importantly still, closely linked to the question of colonisation. Botany is both a referential element in the novel and a metaphor. As a metaphor, it stands for an imperialistic desire to reduce and classify the other in a preconceived nomenclature which is not only artificial and reductive but also quite coercive in imposing the domination of Western science and cultural approach. Two visions are being contrasted in the novel: an idealistic desire on the part of Mr Frazer, a local minister and botanist, to use botany as a way of getting to know the continent and its endemic species so as to make significant scientific progress in the European knowledge of plants and their virtues; and a temptation, on the other hand, on the part of colonial authorities, to use language merely as a means of coercion, control and appropriation of the land and its various specimens, whether botanical, animal or human.

Science and Empire work hand in hand

Remembering Babylon takes place in the middle of the 19th century, at a time when Linnaeus's system of classification was still being widely used, even though it betrayed an approach inherited from the 18th century in which botanical specimens were described in a static form, as a set of characteristics:

> As Linnaeus says, the naturalist–whom he calls *Historiens naturalis*– 'distinguishes the parts of natural bodies with his eyes, describes them appropriately according to their number, form, position, and proportion, and he names them.' (Foucault 2002, 175)

The emphasis is on the sense of sight and this is why sketches are so important to the amateur botanist in the book, namely Mr Frazer:

> [...] he sketched the parts of the plants Gemmy showed him, roots, leaves, blossoms, with straight little arrows in flight towards them from one side or other of the page, where Mr Frazer, in his careful hand, after a good deal of trying this sound and that, wrote the names he provided. (Malouf 1994, 66)

In Linnaean classification, plants are described independently of one another and only linked according to similarities concerning a limited number of physical characteristics. The idea of associating plants according to more fundamental characteristics pertaining to their similarities as living organisms only appeared at the beginning of the 19th century. To that extent, Mr Frazer is still very much a man of the 18th century, for whom botanising is above all a question of making an inventory of the world's biodiversity, more than an attempt to elucidate plant physiology:

> The 18th century ends with the thriving of methods which permit to name, classify and describe animal and plant species, whether animal or vegetal, to work at last on this "great catalogue of existing beings" that one is tempted to call retrospectively the inventory of biodiversity. (Drouin 2008, 50, my translation)

Mr Frazer is a conscientious collector of botanical information about such biodiversity, which is so complex that it can only be recorded according to a set of meticulous steps such as accurate drawings with accompanying notes and entries describing the main parts of the plant. But such a highly codified way of recording the "complexity of things" (Malouf 1994, 129) implies a form of relative blindness.

And such a blinkered vision of his environment is made all the clearer when Mr Frazer, who also happens to be the local minister, goes botanising with Gemmy, a white man who has been very much influenced by aboriginal culture since he spent 16 years with a native tribe as the sole survivor of a shipwreck off the Australian coast. As a "black white man" in the eyes of the settlers (Malouf 1994, 10), he is of great help to Mr Frazer's botanical investigations. There is a lot the minister simply does not see and it is up to Gemmy to enlighten him or to leave him in obscurity and ignorance. So, whereas Gemmy is "moving through a world that [is] alive for him and dazzling," Mr Frazer "[sees] nothing at all" (67-68). Mr Frazer goes "barging through" and Gemmy does not "enlighten him." David Malouf, in contrasting two ways of apprehending botany and more generally the world, invites the reader to question his assumptions about the superiority of the Western approach to his environment: an approach which is very systematic and rational but which also leaves a lot in the dark. By reversing the cliché metaphors about enlightened colonisers bringing the torch of civilisation and Western knowledge to people living in the darkness of ignorance and backwardness, he points to the question of epistemological blindness. The Western scientific and systematic approach paradoxically implies blind spots, unseen patches, just as Buffon's and Linnaeus's scientific approach in their time induced a concomitant blindness. As Foucault says:

> Buffon and Linnaeus employ the same grid; their gaze occupies the same surface of contact upon things; there are the same black squares left to accommodate the invisible; the same open and distinct spaces to accommodate words. (Foucault 2002, 148)

Frazer is partly conscious of this problem when he underlines the paradox of early settlers starving where their Aboriginal counterparts would survive easily:

> I think of our early settlers, starving on these shores in the midst of plenty they did not recognise, in a blessed nature of flesh, fowl, fruit that was all around them and which they could not, with their English eyes, perceive, since the very habit and faculty that makes apprehensible to us what is known and expected dulls our sensitivity to other forms, even the most obvious. (Malouf 1994, 130)

Their European mindset and horizon of expectations leads to a form of epistemological blindness. But Mr Frazer is less clear-sighted about his own limitations. He does not realise for instance that his limited vision and his highly specific grid of observation inherited from Linneaus prevent him from seeing life in its complex growth: for Gemmy, on the contrary, the world is "crackling and creaking and swelling and bursting with growth" but he "casts the light only in patches for Mr Frazer, leaving the rest undisclosed" (Malouf 1994, 68). To that extent the minister is a typical 18th century naturalist, a man concerned with the world of plants in its "structured visibility" as Foucault says: "The naturalist is the man concerned with the structure of the visible world and its denomination according to characters. Not with life." (2002, 176)

So, even though Mr Frazer sounds sincere in his quest for further scientific knowledge, his work is strongly put at a distance by his cultural and epistemological limitations. He also fails to see what British colonisation is mainly about: mercantilism, a utilitarian use of science and taxonomy.

In *Remembering Babylon*, naming, classifying, whether plants or animals, is about establishing a set of distinctive features that allow the coloniser to impose an order and names on things and men that establish his omnipotence and a grid which serves his own colonial interests, far from any scientific desire of a greater understanding of the mechanisms of life. There is a reflection on the role of naming that runs through the whole novel and that is not only restricted to the question of botanical specimens.

It is comic enough for Frazer to misspell and mispronounce indigenous words which then refer to something completely different but it can also be sacrilegious as when he asks Gemmy for the indigenous name of certain plants which only women are supposed to deal with. More significantly still, Frazer seems totally unaware of the possible repercussions of his own botanising work. With such a systematic approach and such an unquestioned prerogative given to language and naming, he does not seem to realise he also perpetuates a system in which Western naming is an equivalent of Western appropriation.

Science as a Means to Subjugate the Colony

If one thinks more generally of the way European settlers have always appropriated the land they colonised through naming, one soon realises that such a systematic approach is very convenient politically speaking. The British theory of a "terra nullius" in need of British map-making was a way to affirm British ownership of the land. As Paul Carter cogently showed in *The Road to Botany Bay*, place-naming is a question of legitimating the appropriation of a land: "the primary object is not to understand or to interpret: it is to legitimate." (1988, XVI) And the anonymous narrator reminds the reader of such political stakes which link naming and map-drawing to the establishment of artificial boundaries which are supposed to attribute a certain area to white settlers, and to fence off unwelcome natives "traipsing this way and that all over the map," "encroaching on boundaries" registered in the Lands Office in Brisbane:

> six hundred miles away, in the Lands Office in Brisbane, this bit of country had a name set against it on a numbered document, and a line drawn that was empowered with all the authority of the Law. (9)

The authority of the Lands Office, just like that of Mr Frazer, is taken for granted by the settlers, as if naming were necessarily a white man's business. The extent to which naming is considered as perfectly natural and legitimate is illustrated by a young boy's fantasies about exploration and place-naming. Reproducing the adult British settlers' arrogance in appropriating the land, twelve-year-old Lachlan imagines himself giving names to unknown places that he would have discovered:

> As soon as he was old enough, with Gemmy as his guide, he would get up an expedition to search for Dr Leichhardt.
> Somewhere along the way he might be wounded by blacks. Gemmy would nurse him back to health with herbs only the natives knew of. He would discover two or three rivers, which he would name after some of his acquaintances, and a mountain to which he might give his aunt's name, Mount Ellen, or the name of some place in Scotland. (Malouf 1994, 60)

Using a young and naïve boy's limited perspective is quite astute on the part of David Malouf. It sheds light on a widely shared conviction and delusion that exploration and place-naming are innocuous and that the few Blacks that the white man might thus encounter are negligible: the wounds they would inflict on the white explorer would soon be forgotten and forgiven in a process of "peaceful" assimilation. For the young boy, there is no doubt that the land is his as long as he decides it to be. Native people are simply anecdotal, peripheral. As soon as the map is covered in white men's and women's names, they disappear from the picture. This is why when Gemmy, a former British subject turned "native," suddenly appears on a picket fence, the young boy's first reaction is to pretend he has a gun and is ready to use it to defend his territory as a white settler. Being a young boy instead of an adult, Lachlan's violence and aggressive assertiveness are both more exacerbated and less obnoxious: such a choice of focaliser reflects an authorial strategy which consists in charming the reader without sounding too critical or too moralizing while at the same time magnifying the violence that naming and the establishment of boundaries really entail.

Similarly, Mr Frazer as a botanist, in trying to fit new botanical specimens into preexisting categories, is also betraying his own systematic way of thinking which consists in drawing artificial boundaries that mostly reflect his own cultural worldview:

> He turns back a moment to his notebook, and what he sees is no longer a wild place but orchards in which, arranging themselves in rows, wild plum and fig and apple have moved into the world of cultivation, and in the early morning light, workers with the sun on their backs hang from ladders and reach out to pluck them. (Malouf 1994, 132)

Here the botanical specimens have lost their singularity and are artificially integrated into rows in an orchard that is typically European. The end of the quotation sounds more like the description of a painting than the meticulous description of a botanist.

But what is even more striking is that the white settlers also want to dispose of the Aborigines in the same manner: instead of granting them any free will, they have decided in advance what place they could have in their "settled space," that of labourers and house-servants, namely menial tasks and functions subordinate to preexisting categories such as landlord/labourer, master/servant, in pretty rows:

> What they looked forward to was a settled space in which they could get on with the hard task of founding a home, and maybe, if they were lucky, a town where in time all the civilities would prevail. If they got the preliminaries right, the natives too might be drawn in, as labourers, or house-servants. They had, secretly, some of them, a vision of plantations with black figures moving in rows down a field, a compound with neat whitewashed huts, a hallway, all polished wood, with an old greyhaired black saying 'Yessir', and preparing to pull off their boots. (Malouf 1994, 62)

Here the vision is less idealised and less poetic: instead of Frazer's vision of noble labourers reaping the rewards of their rustic work, we have a vision of the exploitation of the black man by the white man with very similar images: the motif of the rows and of cultivation. The fields have replaced the orchards but more importantly still, the interrelation between cultivation, colonisation and exploitation is made explicit. As Robert

Young indicates in his book *Colonial Desire: Hybridity in Theory, Culture and Race*, Culture in its botanical sense of cultivating a plant is indistinguishable from its imperialistically oriented political sense of colonising and appropriating a land and its inhabitants:

> 'Culture' comes from the Latin cultura and colere, which had a range of meanings: inhabit, cultivate, attend, protect, honour with worship. These meanings then separated out: with Christianity, the 'honour with worship' meaning of cultura became the Latin cultus, from which we derive our word 'cult'–and from which the French derive their word couture. More significantly, the 'inhabit' meaning became the Latin colonus, farmer, from which we derive the word 'colony'–so, we could say, colonisation rests at the heart of culture, or culture always involves a form of colonisation, even in relation to its conventional meaning as the tilling of the soil. The culture of land has always been, in fact, the primary form of colonisation; the focus on soil emphasizes the physicality of the territory that is coveted, occupied, cultivated, turned into plantations and made unsuitable for indigenous nomadic tribes. (Young 1995, 30-31)

And this is a question of which Mr Frazer does not seem to be fully aware. Whereas his use of botany tends to only equate scientific progress with classification and nomenclature, the natives' approach appears much more respectful of nature itself. For Frazer, nature is to be dominated by man, man's observation, naming and ultimate tilling of the land so as to extract as much profit from it as possible. The minister illustrates here a widely shared conviction at the time, that of the necessity for science and empire to work hand in hand to promote and further British interests. As Gascoigne shows in his book *Science in the Service of Empire*, there was a doctrine which summed up such a policy—the doctrine of *improvement*. It provided a moral veneer to justify colonisation and reap as much profit as possible for the metropolis under the guise of universal scientific progress (Gascoigne 1998, 166-170). For the Aborigines on the other hand, man should adapt to nature instead of trying to *improve* it.

The Empire writes back to Western science

Botany is a scientific field which is typically Western but in fact Aborigines could provide a very fruitful complementary approach, maybe less focused on formal characteristics and more on practical uses and the two are highly complementary. Another point of interest is that the Western use of botany tends to try and set up universalising systems whereas Aboriginal knowledge is more centred on local and cultural characteristics.

Western botany strives for exhaustiveness and universality while Aboriginal knowledge divides its knowledge and its handling of plants into masculine and feminine knowledge, neutral and taboo.[32] So, if Frazer really wanted to take down Aboriginal names, he should only take down the names of the plants that men are allowed to deal with and not others. This is where we see that the knowledge of plants, Western knowledge with its scientific approach as exemplified in botany on the one hand, Aboriginal knowledge with its wider cultural approach on the other, contrasts two world views: one that takes for granted that nature is there to serve human interests in its illimitable bounty and another that sees man as subservient to nature and therefore in need of worshipping its sacred dimension. There is the question of changing man rather than nature, changing taxonomies and "culture," a way to inhabit and cultivate the world so as to adapt to an environment that imposes its own rules and constraints, its own order or disorder, its configuration, its "tumbling complexity" to use Mr Frazer's words.

The best example of such arrogance on the part of Bristish settlers is to be found at the end of the novel, when Mr Frazer convinces the Governor to receive him. Hoping that his botanical report on native plant specimens will be put to good use in a typical utilitarian way, in developing the extensive and systematic cultivation of the most profitable fruit or vegetable species, Mr Frazer presents his report to Mr Bowen. But the Governor turns out to be a very pompous and suspicious man who is at first afraid there might be some plot against him in the report. When he realises Mr Frazer is harmless, he simply loses interest. Such an attitude may point to the relative amateurism in colonial administration in the first decade of the 19th century. And when Mr Frazer then meets the Premier, he is faced with a similar indifference. The latter, Mr Herbert, seems to be so much engrossed in reproducing a miniature England in Australia that his interest in botany restricts itself to the matter of acclimatising European plant specimens in Australia rather than developing the cultivation of native plants. When he arrives at the Governor's place to meet Mr Frazer, he comes with a basketful of vegetables from his own garden. His estate, Herston, is only meant to be an almost identical replica of a British estate: "the fifty acres of Cambridgeshire he has established in a place, that, once he leaves it, he will not revisit" (Malouf 1994, 173). He clearly sees himself as a visitor who does not seem to want to improve Australia for itself but simply to make his stay as pleasant as possible.

In his garden he has managed to grow asparagus, strawberries but also far more exotic plants and animals: "from grapes and China peaches to its mouse deer, its Breton cows, its Arabian bull, its peacocks, pheasants, guinea pigs" (174). So, on a small and amateurish plane, Herbert acclimatizes plants and animals from all over the world for his own benefit and pleasure but shows no interest whatsoever for endemic species. Not once does he enquire about native flora and fauna: he does not even ask Mr Frazer any question on his botanical report as if to suggest how little he cares about it. To that extent, he perfectly illustrates the growth of acclimatisation throughout the 19th century, a cultural phenomenon which was not so much interested in the progress of universal scientific knowledge but really concerned with questions of its utilitarian uses: how to improve the settlers' way of life:

> [...] the acclimatized exotic organism functioned as a symbol of Europe's power over nature and over far-off lands. Vested with visions of imperial superiority, acclimatized animals and plants provided material manifestations of science serving the interest of transplanted Europeans. Relying on exotic plants and animals, acclimatization schemes also tended to devalue indigenous methods of agriculture, and probably degraded colonial environments. (MacLeod 2000, 151)

The Premier, Mr Herbert, probably saw no obvious opportunity in the study of Frazer's report for furthering British interests and therefore simply ignored the subject.

Scientific classification, map-making and political control

Significantly enough, Mr Frazer's report ends up, not with its botanical expertise being recognised and used to experiment on "orchards" of native fruit trees but as a pretext to try and integrate Gemmy into the artificial bureaucratic colonial arborescence. So instead of furthering the scientific knowledge of the place, the report is crushed in the purely political and administrative cogs of a ruthless and blindfolded imperialist machine, which has no interest whatsoever in native life, whether human or natural, except when it can further its own appropriation and expansion policies. Gemmy is ironically given a position as "customs officer" (Malouf 1994, 175), thus suggesting his main function will be to make sure everybody respects a system of artificial boundaries based on unfair appropriation of the land. Indeed, from a purely legal standpoint, there is no reason why the British settlers should claim such a right on the land.

Another ironical touch is that the capital of such an outpost of the British Empire, namely Brisbane, should be described precisely as a grid of interlocking "picket fences" (Malouf 1994, 175), the very term used in the first scene when Gemmy faces the children of the colonisers: empire-building is clearly assimilated with the artificial imposition of a preordained map, an imperial hierarchy which does not try to learn from the place it conquers but only to impose an exact copy of the original. To that extent, it is emblematic of the *bureaucratic* way colonial administration works, its tendency to transpose an exact copy of its own hierarchical structure: as Deleuze and Guattari say, "Accounting and bureaucracy proceed by tracings" (2004, 16). And instead of such a blind and narrow-minded reproduction of earlier structures, colonial administration could have used Gemmy's expertise on Australian botany and Aboriginal culture. The Premier and the Governor could have put to good use his hybrid status as a former British subject turned native for a while. His in-betweenness defying any form of nomenclature or classification could have helped them escape arborescent thinking based on a reproduction of the same and they might have ventured instead into rhizomatic thinking as a way to question their own worldview:

> We're tired of trees. We should stop believing in trees, roots, and radicles. They've made us suffer too much. All of arborescent culture is founded on them, from biology to linguistics. Nothing is beautiful or loving or political aside from underground stems and aerial roots, adventitious growths and rhizomes. (Deleuze and Guattari 2004, 16)

Botany in *Remembering Babylon* is therefore a very illuminating illustration of the complex interaction between science and empire in the 19th century. Even if the discipline was to evolve radically all along the century from a descriptive scientific approach based on close observation of a set of specific characters inherited from Linnaeus to a more dynamic and fundamental approach centred on questions of plant physiology and development, such a reappraisal of the notion of life and plant specimens as living organisms and not only aggregates of distinct parts was less easy concerning human specimens.

Throughout Australian history, Aboriginals have also been classified according to a series of physical criteria such as the colour of their skin which had nothing to do with their identity and their culture. But it is obviously much more convenient to dispose of people when they are turned into mere statistics or measurements than reassess one's own culture and interpretative tools when faced with another culture. With the stolen generations and the assimilation policy aiming at breeding out aboriginality and wiping out their culture well into the mid 20th century, one can see that such epistemological revolution concerning men has taken a very long time indeed.

Works cited

Bourguet Marie-Noëlle and Christophe Bonneuil. 1999. Dossier thématique: De l'Inventaire du monde à la mise en valeur du globe: Botanique et colonisation. In *Revue Française d'Histoire d'Outre-mer* 86 : 322-323.

Brantlinger, Patrick. 2003. *Dark Vanishings. Discourse on the Extinction of Primitive Races, 1800-1930*. New York: Cornell UP.

Carter, Paul. 1988. *The Road to Botany Bay. An Exploration of Landscape and History*. New York: Alfred A. Knopf.

Deleuze, Gilles and Félix Guattari. 2004. *A Thousand Plateaus*. Trans. Brian Massumi. London and New York: Continuum. Vol. 2 of *Capitalism and Schizophrenia*. 2 vols. 1972-1980. Trans. of *Mille Plateaux*. Paris: Les Editions de Minuit.

Drouin, Jean-Marc. 2008. *L'herbier des philosophes*. Paris: Seuil.

Foucault, Michel. 2002. *The Order of Things : An Archeology of the Human Sciences (Les mots et les choses)*. Routledge.

Gascoigne, John. 1998. *Science in the Service of Empire: Joseph Banks, the British State and the Uses of Science in the Age of Revolution*. Cambridge: Cambridge UP.

Glowczewski, Barbara. 1991. *Du rêve à la loi chez les Aborigènes. Mythes, rites et organisation sociale en Australie*. Paris: PUF/Ethnologies.

MacLeod, Roy, ed. 2000. *Nature and Empire: Science and the Colonial Enterprise, Osiris. A Research Journal Devoted to the History of Science and its Cultural Influences*. Second Series. Vol.15.

Malouf, David. 1994. *Remembering Babylon*. New York: Vintage.

Osborne, Michael A. 2000. Acclimatizing the World: A History of the Paradigmatic Colonial Science. In *Nature and Empire: Science and the Colonial Enterprise, Osiris. A Research Journal Devoted to the History of Science and its Cultural Influences,* ed. Roy MacLeod. Second Series. Vol.15. 135-151.

Street, Brian V. 1975. *The Savage in Literature, Representations of 'primitive' Society in English Fiction 1858-1920*. London: Routledge.

Young, Robert. 1995. *Colonial Desire: Hybridity in Theory, Culture and Race*. London: Routledge.

THE DISCOURSE OF EMPIRICISM
AND THE LEGITIMIZATION OF EMPIRE

ANNE LE GUELLEC

Jeremy Bentham was dismayed when, in 1788, deportation to Australia was preferred to the Panopticon, his scheme for an ideal prison, as an answer to an accumulating convict population who could no longer be sent to the American colonies. Strongly protesting against the penal colony of New South Wales, he conceded sarcastically in 1812 that transportation

> was indeed a measure of *experiment* [...] but the subject-matter of experiment was, in this case, a peculiarly commodious one; a set of *animae viles*, a sort of excrementitious mass, that could be projected, and accordingly was projected [...] as far out of sight as possible." (Bentham 1776, 7)

What was perceived as a definitive and profitless removal of a criminal population far beyond the scope of enlightened supervision could indeed scarcely be considered as a valid "experiment" to Bentham who believed that "correspondent to *discovery* and *improvement* in the natural world, [was] *reformation* in the moral." (1776) Even before it became tainted by the penitentiary beginnings of its colonisation, the Australian continent held little promise of profit for imperial Britain and fantasies of weird or hellish antipodes would have made settlement so far away from Britain appear at best tentative. Later, when the colonisation of Australia turned out to be successful, the natural surroundings having proved less lethally alien than expected, the word "experiment" was still consistently used to refer to this outpost of Empire in the Pacific, particularly in the context of the evolutionary and racial debates of the 19th and early 20th century. This paper will briefly trace the history of these "experiments" by which science, and in particular medical science, established itself as an authority on what the destiny of Australia should be within Empire. It will also attempt to explore how the empirical and shifting definition of whiteness that was science's main concern in Australia, both legitimizing colonisation and unsettling it, may be responsible for creating a paranoid sense of [national] identity which is diversely reflected in colonial and post-colonial literature.

The philosophical debate between the Continental rationalists and the British empiricists in the 17th and 18th centuries founded the convenient if oversimplified contrast between a continental approach traditionally defined as favoring reason and theory and the more sceptical British approach relying on sensation and experience. Certainly, the empiricist legacy seems particularly clear in the way the colonisation of Australia was consistently represented as an experiment in the British imperialist discourse of the 19th and early 20th centuries. Contrary to the French revolutionary imperial discourse, or to the concept of Manifest Destiny in the independent American colonies, the colonizing and even proto-national discourse of Australia always includes failure as a possible, and indeed very often as a *probable* outcome. This is particularly true in the field of medicine, which Warwick Anderson, an Australian medical doctor and historian, has documented and analyzed in his fascinating book *The Cultivation of Whiteness* (2006). In his introductory pages, Anderson justifies his focus on medicine as a way to contribute to a more complete history of imperialism in Australia by stating that:

> it is important to realize that medicine, until the early twentieth century, was as much a discourse of settlement as it was a means of knowing and mastering disease. In seeking to promote health, doctors drew on a fundamentally moral understanding of how to inhabit a place with propriety. In Australia, most doctors assumed that only whites would ever reach the necessary standard of hygiene and decorum; some of them wondered if there were places that were inimical to a cleanly, self-possessed, white civilisation, places that could never be successfully colonized and remain purely white. (Anderson 2006, 4)

The perceived strangeness (both mythical and real) of the natural environment in the antipodes partly accounted for the pessimism of the first colonists, but so did the contemporary belief that each race had a distinctive constitutional character or temperament that was best suited, whether through providential or evolutionary mechanisms, to its ancestral environment. It was therefore expected that the first British immigrants would be unable to cope with a place in which nature was so different from that of their native land unless they were able to acquire new knowledge about it. The chances of survival depended therefore on the adaptability, both physiological, but also cognitive, of the British race. In this last regard, colonial doctors were to stress the vital importance of the extensive gathering of clinical data, but also to claim that if they were to practice with sensitivity, they had to emancipate themselves, as Andrew

Ross, in an article entitled "The climate of Australia viewed in relation to health" put it, from "quixotic notions of the universality of medicine" and to develop instead a medical science "from an Australian point of view" (Anderson 2006, 35).

By the 1850s, the adaptability of the European race was confirmed by the expansion of the colonies, but it itself proved a source of somber speculation, as the current Lamarckian theories of adaptive heredity drove doctors to seek evidence of the degeneration of the virile, intelligent and civilized white race under the unfavorable influence of the austral environment—particularly in the tropics—which had produced the unprepossessing Aborigine. For example, Dr James Kilgour in the 1850s commented on the "slighter muscular system of the Australian youth" known as a result as the "cornstalks," stating that he had observed the same puny build in local Aborigines, horses and cattle (Anderson 2006, 27).

However, after the publication of Darwin's *Descent of Man* in 1871, which held that human heredity in civilized communities was more or less fixed and autonomous regarding new environments and alteration in habit, racial degeneration began to be seen less as the result of a good adaptation to bad conditions, than as the result of an overprotective civilisation preventing the environment, whatever its qualities, from weeding out the unfit. This shift meant that if displacement was no longer responsible for the degeneration of the stock, it might actually be helpful in its regeneration, in combination with an improved supervision of social processes, in particular reproduction. Indeed, the scarcity of the fast-disappearing black race and the very isolation of the island-continent, which made it possible to control immigration from Asia, came to be seen as assets in the great colonizing experiment of tropical Australia. As Reverend Dr Frodsham wrote in his recollections published in 1915:

> The whole question of the adaptability of race and environment is too obscure to allow dogmatism with regard to the suitability of the tropics to a white race. [...] When all is said and done, the whole thing [colonisation] is only a big experiment. It is an experiment worth trying. Australia has seen the vision of keeping a home for the white race in the southern seas. It is seeking to guard the existence of white civilisation from being crowded out by a lower social organism. Australians desire to guard the purity of the white race. (1915, 101)

Towards the end of the 19th century therefore, against a background of high imperialism, which defined economic and political rivalries as a race struggle, doctors no longer perceived the Australian experiment as the unnatural and doomed transplantation of British population sundering bonds of blood and soil, but as a laboratory where a new white race might be created.

It can be said, therefore, that in spite of repeated claims of epistemological freedom from European doctrines, colonial medicine in Australia was quite typically less empirical than ideological, actual experience doing very little to dispel anthropological and cultural prejudices. Indeed, very basic data indicating, for example, that the European metabolism was not radically altered by the climate, or at least not in any way differently from the non-white metabolism, did not discourage doctors from giving "objective," biological interpretations for the probable psychological suffering of dislocated populations. The specificity of the "Australian experiment" therefore lies elsewhere. As Anderson shows in his book, the real legacy of medical research in Australia is

> not so much the establishment [...] of a medical specialty in tropical medicine as the permeation through all levels of society of an assumption that most political problems—tropical or otherwise—have biological causes and solutions and that medical scientists were social experts. (Anderson 2006, 186)

In the late 19th and early 20th centuries, the medical elite were very active in the major Australian cities, particularly in Melbourne, and by deploying statistics, by vulgarizing their findings and organizing campaigns of information on public health, they were able to translate the complex and messy political process of nation-building into a clear scientific idiom, and to help turn a colony with no great founding myth into a new nation provided with a sense of destiny in which it could take pride. As the Melbourne political economist G. L. Wood put it in 1926, to have the problem of population policy "lifted from the arena of party politics to the laboratory of the scientist is a very real contribution to an ultimate and successful solution of [Australia's] greatest national problem." (Wood 1926, 232)

In Australian intellectual circles there were of course some dissenting voices such as Bernard O'Dowd, the poet and socialist who, in 1912, delivered a speech on the "unscientific" nature of race prejudice before an audience of Victorian theosophists. He reminded his listeners that it was only in "comparatively recent times that men have set out deliberately to give us scientific reasons why we should hate our brothers of the coloured and other races" and warned them against the "extremely subtle and plausible and dangerous pseudo-science of Eugenics" (O'Dowd 1912, 155, 175). But if it can be said that a sense of insecurity was the main legacy of the colonial history of the transplantation of a small European population to a vast and impossibly distant land, then it is hardly surprising that the crystallization of this passive anxiety of dislocation into a militant if vague concept of whiteness would have made the "White Policy" more generally acceptable than any other. The paradox is that medical science had helped increase the bewilderment of the "white Australian" who, as time went by and immigration diversified, could no longer define himself in reference to British lineage nor to European origins, nor even to the color of his skin. For the new Australian, since the discovery that Aborigines had the same blood groups as Caucasians, and could therefore be homeopathically added to the other acceptable stocks, was expected to be relatively swarthy. The definition of the white civilized Australian could not be based on strictly biological criteria such as blood, morphology, physiology and heredity but involved criteria of behavior, customs and culture, which were nevertheless all subsumed under the scientist's approach. The scientist's experimental unsettling of whiteness also guaranteed that he was the only expert capable of resettling it, and legitimizing colonisation in the process. As a consequence, a pseudo-scientific vocabulary was commonly used to express wide-ranging concerns about a possible loss of civilisation incurred by the dislocated coloniser.

The question concerning Australian literary mediocrity which tormented the intellectual elite until the 1950s, was put in characteristically technical terms in an 1856 essay by Frederick Sinnett, entitled "The Fiction Fields of Australia" and considered to be the first substantial critical work on Australian literature. Sinnett justifies his "inquiry into the feasibility of obtaining the material for this sort of manufacture [the novel] from Australian soil" by stating that it is in the shape of novels that "civilised man, at the present day, receives the greater part of the fictitious clothing necessary to cover the nakedness of his mind." (1997, 19). Here the opening reference to the manufacturing and mining industries which anthropologically distinguish the technologically-advanced white man from the backward black native is rather unexpected in the inquiry into the possible emergence of an Australian literature, but in the same vein, Sinnett sets the reader's anxieties at rest at the end of the essay by a reference this time to the "mathematical doctrine of probabilities" as applied to genetics:

> If only now and then out of the population of all England there arises a novelist capable of breaking up fresh ground, it is not to be wondered at that no such man has yet arisen here. Geniuses are like tortoiseshell tom-cats—not impossible, only rare. Every ten years one is born unto Great Britain, but probably none exists in Australia, and a reason precisely

analogous to this makes it improbable that we have at present among us any one capable of doing justice to Australian materials of fiction. There are not enough cats in Australia to entitle us to a tortoiseshell tom yet, according to the doctrine of averages. (Sinnett 1997, 23)

In these closing lines, the reassuring scientific analogy conveniently explains away the absence of good Australian fiction in the 1850s, in an attempt to relieve fears of colonial insufficiency, itself being a possible symptom of the failure of the white race in the antipodes. Conversely, one of the conditions for the white man to succeed in complete legitimacy, namely, the disappearance of the black man, was often presented in both scientific and lay texts with a strangely unscientific inconsistency. One such text published in 1928, is a descriptive piece by the Victorian poet Mary Fullerton entitled *The Australian Bush*. In the chapter devoted to "The Aborigines?" the first few lines perfunctorily introduce the "controversy" concerning their origin, whether Caucasian or African, but this is dismissed as not being a very important issue:

Wherever he came from, when the white man took possession of the Australian continent, the blacks ranged over the whole country in tribes, each within its defined boundaries. Even then they were a dying race, especially in the southern part of the continent. Tasmania lost its last poor remnant long ago, and Victoria has but fifty-five full-blooded blacks left of its various tribes. (61)

After having thus complacently dissociated the white man's takeover of the continent from the Aborigines' disappearance, Fullerton moves on to the northern parts of Australia where the estimated number of natives is said to be somewhere between fifty and two hundred and fifty thousand. But just as dismissively as in the first paragraph, Fullerton moves beyond the scientific uncertainty to assert: "In any case they are rapidly decreasing. Civilisation disagrees with this primitive people." Having given one example of how civilisation may have harmed the Aborigine (unused to wearing clothes, he is the victim of consumption by not taking them off when they get wet), Fullerton then refers to the Aborigine's "spasmodic, and almost futile" attempts to fight against the "white invader," and concludes this introduction with a sorry portrait of the strangely hopeless Aborigine:

As a menace to settlement [...] the native is practically dead. He will linger on, a wretched remnant to be cared for by the whites, for fifty years or so yet; when he must disappear, leaving hardly a mark of his poor vagrant existence on the face of the land. (61)

With the fate of the Aborigine thus sealed, Fullerton moves on to a cursory description of his customs, beliefs and hunting techniques. But she keeps returning to the strange if inevitable fact of his extinction, as if she were at the same time reassured and unsettled by it:

> In a land subject to drought, and where natural food is not over-plentiful, the dying-out in times of stress of an unintelligent race is inevitable. [...] One reason alleged for the aborigine's failure to make provision for lean times while food is plentiful is that he has a superstitious belief that by certain magic arts he can create a supply at need. It is curious that experience has failed to cure him of this superstition. (64)

It seems that what most strikes Fullerton in the Aborigine's fate is his incapacity to acquire knowledge through experience and to adapt to his surroundings, and this of course in spite of having ranged all over the Australian continent for some time before the arrival of the white man. It may be argued that this text is not Fullerton's best work but a rather careless piece of anthropological popularization, yet it is striking to note that the dying-out of native Australians is similarly presented as a natural or even mystical phenomenon in contemporary scientific accounts. The specific anatomy of the Aborigine and his origins were precisely documented and widely speculated upon, but the pathologies which led to his disappearance did not seem to be the concern of the scientist. Even when bacteriology progressed, "civilisation" was still called upon to vaguely account for his mysterious "dying-out."

Possibly this refusal to recognize the coloniser's responsibility in the active and accidental extermination of the native Australians could be accounted for by a suppressed sense of guilt; possibly it was cynically more convenient to pretend this disappearance was natural rather than seek economic or social solutions to a problem, which was, in effect, the solution natural selection was bringing to reinforce the White Policy.

But a striking parallel may also be established between the figure of the doomed white Australian, and that of the doomed black one. Strangely echoing Fullerton's lack of logical consistency in the piece quoted above, the novelist Marcus Clarke wryly predicted that a century from 1877, the average Australian male would be "a tall, coarse, strong-jawed, greedy, pushing, talented man, excelling in swimming and horsemanship," but that within five hundred years, "the breed [would] be wholly extinct" (1877, 20, 22). As has been said before, medical records and lay articles predicted until quite late in the 19th century that the white man was doomed in Australia, in spite of contrary evidence. It might therefore be argued that white Australians remained intimately insecure regarding the legitimacy of their own historically belated appropriation of the southern continent and were projecting their anxieties onto earlier Australians, whom science had also presented as Caucasian invaders, and that the unquestioning and repetitive reference to their inevitable disappearance was the indirect expression of their own insecure identity.

Interestingly, but hardly unexpectedly, post-colonial fiction reverses the colonial perspective by focusing less systematically on the extinction of the Aborigine than on the fantasized, mysterious vanishing of the white man. One may think of such popular fiction as Peter Weir's *The Last Wave* or even *Picnic At Hanging Rock* in which the Aborigine, incarnate or symbolically represented by elements of topography, is used to herald the violent or acquiescent annihilation of the white protagonist. Among Australian novels, David Malouf's works of historical fiction are particularly haunted by the frailty of the white man's presence when confronted by an aboriginal Other. In *Remembering Babylon*, for instance, a rather intrusive narrator forces the reader into a confrontation with a black fantastical figure which stands both for the actual Aborigine and for the fears the coloniser experiences when confronted with the image of his own uncivilized origins:

> [T]he horror it carries to you is not just the smell, in your own sweat, of a half-forgotten swamp-world going back deep in both of you, but that for him, as you meet here face to face in the sun, you and all you stand for have not yet appeared over the horizon of the world, so that after a moment all the wealth of it goes dim in you, then is cancelled altogether, and you meet at last in a terrifying equality that strips the last rags of your soul and leaves you so far out on the edge of yourself that your fear now is that you may never get back. (Malouf 1994, 43)

In the historical context of the novel, the peaceful confrontation offered by the hybrid figure of Gemmy is presented as a missed opportunity for reconciliation between the native and the prejudiced coloniser, but Malouf also stresses the violating, dislocating and almost annihilating impact the experience of "terrifying equality" has on the psyche of a settler who is unsettled rather than reassured by the type of relationship science has contributed to establish between the Aborigine and himself, to the exclusion of any other.

As a conclusion, I would now like to return to the question of the links that can be established between science and empire in the case of the colonisation of Australia. In spite of the empirical discourse deployed to account for the different stages of the colonisation of this strange and mythic land, it is clear these rather serve to legitimize a course of action which was driven, not by the disinterested thirst for knowledge, but by the imperial rivalry with other European nations and the increasing need to expand territorial domination. As to the reasons which determined the so-called settlement "experiment," they also seem to be more societal than scientific, as it is in the English aristocratic and liberal tradition—perhaps founded on what Emmanuel Todd has identified as the "atomic" family model which promotes social mobility and reconversion—to resort to the externalization of society's unprofitable members by displacement or transportation, as illustrated in the enclosures movement in the British Isles.

The choice of establishing a faraway colony where the dregs of society might conveniently be disposed of seems therefore to be in full agreement with the way the English society naturally works, but the problem for the colony itself is that this does little to define a colonial destiny with which the immigrant populations might identify. In a way, it could be said that in Australia the scientific discourse and the tentative and pragmatic approach have stood in for an ideologically well-defined colonial project. In this paper, I have attempted, first of all, to show how the scientific discourse has legitimized the colonisation as an experiment in survival; next, as an experiment in adaptation in a competitive global environment; and finally as an experiment in the capacity of the white race to assimilate and ultimately replace any other population.

The question remains of the lasting impact of the Australian scientific experiment. It has certainly profited science itself, as in Australia medicine and anthropology have been free to experiment, but have also been confronted with the failure, the incoherence or the imperfection of evolutionary theories. It cannot be said to have profited the Aborigines themselves, as science did little to protect them against the pathological causes of extinction, but neither was the reification of the Aborigine by the scientific approach worse than the outright extermination cynically legitimated by the greed for land or by demographic pressure as was the case in the colonisation of North America. But the most lasting impact of the scientific discourse may have been on the white Australian imagination, which still seems haunted by an anxious sense of illegitimacy and lack of belonging.

Works cited

Anderson, Warwick. 2006. *The Cultivation of Whiteness: Science, Health, and Racial Destiny in Australia*. Durham, NC: Duke University Press.
Bentham, Jeremy. 1776. *A Fragment On Government*.

http://onlinebooks.library.upenn.edu, accessed Jan 28, 2009.

—. 1812. *The Panopticon versus New South Wales, Two Letters to Lord Pelham*. London: R. Baldwin.

Clarke, Marcus. 1877. *The Future Australian Race*. Melbourne: A. H. Massima.

Frodsham, G. H. 1915. *A Bishop's Pleasaunce*. London: Smith Elder.

Fullerton, Mary. *The Australian Bush*.
    http://purl.library.usyd.edu.au/setis/id/p00026, accessed Jan 28, 2009.

Malouf, David. 1994. *Remembering Babylon*. London, Sydney: Vintage.

O'Dowd, Bernard. 1912. Race Prejudice. *Theosophy in Australasia*. 2 September: 153-6; 1 October: 175-9.

Sinnett, Frederick. 1997. The Fiction Field of Australia (1871). In *The Oxford Book of Australian Essays* ed. Salusinszky, Imre. Oxford, Melbourne: Oxford University Press Australia.

Todd, Emmanuel. 1990. *L'Invention de l'Europe*. Paris: Seuil.

Wood, G. L. 1926. The Immigrant Problem in Australia. *Economic Record* 2: 229-239.

SKELETONS IN THE CUPBOARD: IMPERIAL SCIENCE AND THE COLLECTION AND MUSEUMIZATION OF INDIGENOUS REMAINS

SHEILA COLLINGWOOD-WHITTICK

In 2002, bowing to pressure from the South African government, the French senate voted to return the remains of Saartjie Baartman, more commonly known in Europe by the disparaging nickname "the Hottentot Venus." Baartman's mounted, articulated skeleton along with her skull and teeth had all been housed in the Musée de l'Homme in Paris since her death in 1816. Though the bottled brain and external genitalia of the young Khoikhoi woman—on public display in the museum until at least 1955—had disappeared, her bones were duly de-accessioned and transferred to her native soil (Tobias 2002, 109). This reversion of what had long been regarded as mere "anthropological exhibits" to their human status as bodily remains returning to their place of origin was felt in South Africa to be a symbolic event of profound significance, and the state funeral organised by the government was attended by many thousands of ordinary citizens.

Though the repatriation of Saartjie Baartman's bones is certainly one of the most highly publicised events of its kind, the campaign which enabled such a cathartic outcome is far from unique. Calls for museums to return human remains to contemporary Indigenous communities who claim an affiliation with them have increased exponentially since the 1970s. One of the consequences of such calls has been to bring to light an episode in the history of former imperial nations that the latter would certainly have preferred to remain hidden from view. For what the thousands of human remains still to be found in the museums of the Western world attest to is the practice—common in certain European colonies during the late nineteenth and early twentieth centuries—of exhuming Indigenous bodies, illegally removing them from burial sites, graves, battlefields and other mortuary sites and shipping them to Europe both for scientific investigation and as anthropological exhibits in museum collections.[33]

According to an editorial in the Australian newspaper *The Age*, the senior curator of anthropology at the South Australian Museum, Philip Jones, estimated in 2001 that the number of Aboriginal artefacts held in overseas collections was around 40,000 (n.p.). In 2002 James Morrison of *The Independent* reported that a working group set up for the purpose had found that two thirds of a "random sample" of 150 British museums both local and national held what he referred to as "grisly reminders of the worst excesses of colonialism" (2002: n.p.).

It is with the abusive relationship that imperial science established with the bodies of Indigenous people that the following discussion is concerned. As anthropologist Nancy Scheper-Hughes has pointed out, "The recent critiques of anthropology have released a torrent of institutional and professional self-analysis" (Scheper-Hughes 2001, 18). Drawing on such analyses but also, more generally, on recent socio-cultural studies of empire, I will focus on four main points:

I.   the intersection between the human sciences and imperialism
II.   the connection between osteological research and British white settler colonies
III. the role of museums in disseminating racist perceptions of non-Europeans
IV.   the ethical implications of nineteenth and early-twentieth century scientific conduct in the study of Indigenous bodies

Human sciences and imperialism

The ideological tenor of much of the research carried out on the colonised peoples of the world during the period of Britain's imperial expansion first started to be recognised in the 1960s when anthropologists embarked on the process of critical self-scrutiny to which Scheper-Hughes refers. Writing in 1973, anthropologist Talal Asad spelled out the imbrication of anthropology and imperialism thus:

> Social anthropology emerged as a distinctive discipline at the beginning of the colonial era, [...] became a flourishing academic profession towards its close, [...] [and] throughout this period [...] devoted [its efforts] to a description and analysis—carried out by Europeans, for a European audience—of non European societies dominated by European power (1973, 14-15).

Contrary to the—still prevalent—opinion that scientists operate in a sterile, value-free zone, immune to the dominant beliefs and mind-sets of the day, a variety of socio-cultural influences can be identified as having helped trace the course taken by the human sciences in the nineteenth and early-twentieth centuries. But, I suggest, unless we recognise the Victorian scientist's ideological commitment to the imperial enterprise, it is difficult to fully account either for the bias, or for the virulence that permeated scientific discourse on Indigenous peoples during the period under review.

Certainly, a great deal of the pseudo-scientific knowledge on non-European peoples produced by the fledgling human sciences was due to the unscientific methods of data collection that researchers employed. It is now well-known, for instance, that many of the exponents of theories postulating the biological and cultural inferiority of "primitive" peoples based their conclusions either on second-hand observations taken from the highly subjective travel literature of the period or on scrutinising first-hand the wretched spectacle of natives offered by the human zoos that, for more than eighty years, enjoyed such phenomenal success in the civilised capitals of the Western world (Lorimer 1996, 21; Jackson & Weidman 2004, 16, 79). As Annie Coombes points out, "'armchair' anthropology [...] with a few notable exceptions, lasted in Britain from 1870 to 1920" (1994, 132-3). Flawed methodology is, however, more of a symptom than a cause of the negative image of non-European peoples that the human sciences bequeathed to the world.

"All anthropological knowledge is political in nature" claims Johannes Fabian (1983, 28) and nowhere is that postulate more clearly evidenced than in the century that followed the emergence of anthropology and the related fields of comparative anatomy, anthropometry, craniology and phrenology as scientific disciplines. When Anglo-Celtic settler colonies were being established, the human sciences were above all concerned with demonstrating the sub-humanity of the peoples whose dispossession and/or elimination were the result, not to say the sine qua non, of European settlement. Nineteenth-century anthropology, as Scheper-Hughes argues,

> was built up in the face of colonial genocides, ethnocides, mass killings, population die-outs, and other forms of mass destruction visited on the marginalized peoples whose lives, suffering and deaths have provided [anthropologists] with a living. (Scheper-Hughes 2002, 348)

Towards the end of the century, however, Britain's imperial interests began to evolve, centring more on the development of tropical exploitation colonies in Africa where, unlike the Indigenous peoples of white settler colonies who had been perceived as an obstacle to development (Palmer

2000, 117), the natives represented the kind of cheap and robust labour force on whose exploitation the colonial enterprise would depend.

Anthropology's response to the changed needs of imperialism was to seek to convince the British government that it was a useful discipline, capable of "facilitat[ing] passive consent from subject races" (Coombes 1994, 109). Between 1893 and 1919, more than half of the presidential addresses to the Royal Anthropological Institute focused on "the practical uses to which anthropology could be put in serving the Empire" (Stander 1993, 409), but the ruling classes remained unconvinced of the utility of financing anthropological research. It was not until the late 1920s that funding began to be forthcoming both from the government and from such organisations as the Rockefeller Foundation which had financial interests in Africa (Stander 1993, 415). It was from that point that:

> Anthropologists helped gather the intelligence about native people which allowed these people to be efficiently ruled and exploited in the colonial interest—that is, in the interests of the white settlers, the colonial administrators, the government in the metropolitan country, and the Western capitalist enterprises in Africa (419-20).

Yet the role of science in the imperial project was not just limited to the collection of information that would facilitate the process of colonisation; anthropologists also helped to morally justify British intervention and tutelage by constructing Africans as an intellectually backward and childlike race in need of firm parental guidance and discipline. As Fabian observes, "Aside from the evolutionist figure of the savage there has been no conception more obviously implicated in political and cultural oppression than that of the childlike native" (1983, 62).

One prime example of the use to which colonial societies put anthropological theory is that of Australia where, in the 1920s applied anthropology "had one client—the state" (Cove 1995, 178). Accordingly, the applied research program developed by Sydney University's Anthropology Department was one that "trained missionaries and colonial administrators whose objectives were to assimilate indigenous peoples" (88).

The idea that each of the world's peoples could be accurately positioned on a scale designed to indicate the relative distance of all non-Europeans from the ideal represented by the civilised white did not, of course, originate in nineteenth-century science. It had already been well established in the 18th century by such "founding figures of anthropology" as Linnaeus, Buffon and Blumenbach (Stepan 1982, 2). Nevertheless, it is true to say that the idea acquired an unprecedented resonance and authority in the nineteenth-century. This was in no small part due to the professionalisation of science during that period—a phenomenon which, as Douglas Lorimer explains,

> ... created a community of experts socially selected from the expanding professional middle-class, and confident in the authority of its specialized, scientific knowledge. Among anthropologists, the comparative anatomists, both from their specialized training and from their affiliation with the medical profession, claimed pre-eminent status as scientists (1996, 24).

It was precisely from "the rise of anthropology as a professional domain," argues Coombes, that "certain assumptions about race and racial characteristics ... derived their credibility and longevity" (1994, 109). In contrast with the more hesitant conclusions on race that resulted from the philosophical speculations of eighteenth century gentleman amateurs, the ideas articulated by the new generation of professionals had a greater ring of certainty about them. Proceeding from the premise that "rigorous measurement could guarantee irrefutable precision" (Gould 1981, 74), nineteenth-century practitioners of the human sciences became obsessive measurers of facial angles, cranial capacity, length of limbs, circumference of skulls—in short any and all observable anatomical variations between human groups. One of the direct consequences of this surrender to "the allure of numbers" (Gould 1981, 74) was, as Michael Adas points out, the perceived need for a plethora of extraordinarily specific measuring instruments:

> Technology, in the guise of instruments from simple calipers to Gratton's craniometer, was enlisted in the search for an accurate technique of skull measurement and comparison. Some investigators stressed the need to measure cranial capacity (Samuel Morton used white peppers and then lead gunshot; Friedrich Tiedemann used millet seed); others such as Anders Retzius and George Combe insisted that what mattered was the shape of the head or the proportions of different parts of the skull or brain. The usefulness of Camper's facial angle was debated, and it was gradually replaced by a bewildering variety of new phrenological gauges, including the cephalic index, the nose index, the vertical index, and the cephalo-orbital index. Thousands and then tens of thousands of skulls were measured in innumerable ways. (Adas 1989, 293)

Yet despite the mass and precision of the numbers collected, scientific measurements were, in themselves, no guarantee that the conclusions drawn from them reflected disinterested, value-neutral thinking. On the contrary, as Gould's *The Mismeasure of Man* has clearly demonstrated, many of the racial theories elaborated in the imperial era were a tissue of favourable inconsistencies, shifting criteria, subjective a priori assumptions, specious claims, glaring miscalculations and convenient omissions.

Osteological research and British white settlers

Let us now turn to what Helen MacDonald refers to as that "mid-Victorian mania" of bone-collecting (MacDonald 2006, 96). It goes without saying that research in anthropology, comparative anatomy and other related disciplines/pseudo-sciences would not have been possible without a regular supply of human remains. As MacDonald puts it, "Vast collections of bodies and body parts were necessary for this science" (2006, 87).Yet, it was by no means any human bodies that provided the raw data for early anthropological studies. White settler colonies constituted by far the most prolific source of the osteological specimens supplied to anthropological institutes and museums the world over, with the bodies of Australian Aborigines being much in demand for the evidence they were purported to offer as specimens of humanity at its lowest level. Most sought after of all, however, were Tasmanian skeletons since the indigenous occupants of Van Dieman's Land (later renamed Tasmania) were, as MacDonald explains,

> thought to be distinctively different from those on the Australian mainland, and by mid-century were understood to be on the brink of extinction. It had the effect of turning their bodies into rare collectibles. (2006, 10)

Though in the founding period of the penal colony Aborigines had tended to be represented textually and pictorially as "noble savages," the arrival of large numbers of British immigrants in Australia during the first half of the nineteenth century rapidly put an end to that earlier, comparatively benign view. As soon as Aborigines showed resistance to the settlers' wholesale appropriation of their territories, a revised understanding of the natives began to prevail in the Australian colonies. The only way of dealing with what were now regarded as sub-human savages was, common wisdom dictated, to exterminate them. In the words of historians Chesterman and Galligan, "Indiscriminate killings of Aborigines accompanied their forced removal from any lands considered useful to settlers" (1997, 33). In the state of Western Australia alone, there were, as genocide specialist Colin Tatz points out "hundreds of massacres between settlement and the 1920s, with the last of them, the Forrest River killings, as late as 1926" (1999, 16). The horrifying banality of frontier slaughter is not only recorded in the oral testimony through which the collective memory of Aboriginal peoples is transmitted, it is also inscribed in innumerable written records to be found in the archives of all Australian states.

The question I now want to pose, then, is this: can any link, other than that of chronological coincidence, be said to exist between the genocidal behaviour of settlers and the anthropological theories under discussion? If we look to recent historiography for an answer, the picture that emerges is one of colonial genocide and racial science feeding off one another in white settlement colonies.

As C.D. Rowley explains, "popular pseudo-Darwinism was the best of all sops to frontier consciences" in Australia (1974, 25), for what it did was to explain the disappearance of Indigenous peoples in such a way as to "den[y] the colonisers' role as morally responsible historical agents" (Lorimer 1996, 25). To take the example of Tasmania, frequently cited as a paradigmatic instance of colonial genocide, settlers justified their slaughter of native tribes on the grounds that, since Aborigines were no more than "the connecting link between man and the monkey tribes," it little mattered that they were being wiped out (Turnbull qtd in Cove 1995, 31). In the view of Henry Reynolds, the racial theories propounded by nineteenth century anthropologists were "an essential component of the intellectual backdrop against which relations between Aborigine and European were enacted" (1974-75, 49).

But racial science did not only rationalise colonial violence, it also directly benefited from it. Native corpses were freshly harvested from scenes of massacre while Native burial sites were plundered for Aboriginal remains. Recounting the aftermath of a massacre in Bathurst in 1824, the Reverend Lancelot Threlkeld reported that:

> "Forty-five heads were collected and boiled down for the sake of the skulls. ... the skulls [were then] packed for exportation in a case at Bathurst ready for shipment to accompany the commanding officer on his voyage to England." (Qtd. in Jopson & Stephens 2000, n.p.)

A further example is that cited by historian Robert Hughes who points out that following a massacre of Aborigines at Risdon Cove, Tasmania, in 1804, "the colonial surgeon, Jacob Mountgarrett, prompted by some anthropological whim, salted down a couple of casks of [the Aborigines'] bones and sent them to Sydney" (1996, 414).

In her recent study of colonial genocide Alison Palmer goes so far as to suggest that Aborigines were murdered for the express purpose of providing " 'specimens' for museums"(2000, 46)—a suggestion that seems to be supported by Paul Turnbull's observation that "There was no outcry by scientific figures when, on a number of occasions between the mid-1860s and 1890s, science became the direct beneficiary of murder" (Turnbull 1997 n.p.).[34]

But whether they were merely the incidental by-product of bloody clashes on the frontier or constituted in themselves the "rare collectibles" that settler violence aimed to produce, Indigenous remains were certainly a highly marketable commodity in nineteenth-century scientific circles and collecting them was, in consequence, a lucrative hobby in white settler colonies. In Australia, body-snatching was, claims Turnbull, "a valuable source of cash for travellers, miners, fishermen and other bush-workers in the often uncertain economy of the frontier" (1997, n.p.). At another level, in addition to the specialised collectors and dealers who operated the world over, "administrators, military personnel and missionaries" were also involved in the business of procuring and trading in human remains (Slogget 2005, 2). Indeed, Coombes estimates that

> by about 1904, many colonial administrators, for better or for worse, assumed that collecting ethnographic 'specimens' for British museums, rather than as personal 'trophies', was an intrinsic part of the job. (1994, 132)

Scientists hoping to attract either official recognition or government funding for their research were only too well aware that the "proof" of biological inferiority which the Indigenous body could be made to yield constituted the keystone of the legitimating discourse on which both colonising nations and colonial governments relied. In the metropolis, the retarded development of Indigenous peoples represented an argument for imperial intervention in the backward corners of the earth, in the colonies the benighted mentality of the savage was cited as justification for the inhuman treatment inflicted on Indigenous peoples.[35]

Disseminating racist perceptions of non-Europeans

Moving on to the Museum, an institution that came into its own in the nineteenth century, we find that throughout the imperial era one of the major roles it fulfilled was that of an immensely popular showcase for osteological specimens purporting to demonstrate the simian affinities of Indigenous peoples. Postcolonial critics have described museum ethnography as "an imperial science," (Shelton 2000, 156) and museums as "temples of empire" (Coombes 1994, 109). As Coombes comments, it is now well established that the museum was "an important element in furthering the objectives of the [British] Empire" (126).

Post-Darwinian evolutionary theory had become quite simply "the most prevalent means of displaying ethnographic material" (Coombes 1994, 120).[36] Accordingly, many of the exhibits displayed in Natural History museums from the mid-nineteenth century onwards represented native peoples as being at the lowest level of human evolution. Recalling a visit made in his childhood to the Hall of Man in the American Museum of Natural History, Gould remembers that in the 1920s the museum

> still displayed the characters of human races by linear arrays running from apes to whites. Standard anatomical illustrations, until this generation, depicted a chimp, a Negro, and a white, part by part in that order. (1981, 88)

A mere fifteen years ago, museum curator Edmund Southworth pointed out that

> human material now in UK museums and institutions was originally collected in the last century as part of an effort to substantiate theories of 'sub-human' or 'inferior human' kinship with other animals. (1994, 24)

"Notions of white supremacy and racial superiority, jingoistic slogans for imperialist expansion, and the vision of a dichotomous world divided between the progressive and the backward have all," Adas reminds us, "been rooted in the conclusions drawn by nineteenth-century thinkers" (1989, 153). And nowhere was there a more effective conduit for such thinking than the kind of evolutionary display described by Gould. It was, in fact, in large part thanks to nineteenth-century natural history exhibits that a wide European public was familiarised with the idea of civilisation as a "unilinear, inevitably progressive movement" (Diamond 2004, 18) in the course of which certain racial groups had been left lagging behind while other, more primitive, races were inexorably doomed to extinction. To understand the far-reaching impact that idea has had in Western cultures, it is important to remember how the institution that first disseminated it was (often still is) perceived.

Though, as Mieke Bal points out, there is general agreement among contemporary museologists that "a museum is a discourse, and an exhibition an utterance within that discourse" (qtd. in Phillips 2006, 134), such was not always the case. In the past, Neil Harris reminds us, "Museums were treated not as places where knowledge was disputed or contested but sanctuaries where it was secure" (1995, 1107). In exploiting the didactic role they had been assigned as "custodians of the truth" (1104) to demonstrate white superiority over non-European peoples, museums must be seen to bear a certain responsibility for the tenacity of racist attitudes still in currency today.

Since, as an archaeologist of the period observed, "[a] collection of human skulls is one of the features of a modern anthropological museum" (Murray qtd. in Simpson 1996, 173), museum curators, like their scientist colleagues, were avid consumers of the Indigenous remains that settler colonies produced in such abundance. More particularly, as Cove points out:

> Given the status assigned to Tasmanian Aborigines, specimens from them were defined as central to any serious collection. By 1850 there were nineteen anatomical museums in Britain alone. Tasmanian Aboriginal skeletal remains were found in most of them. (1995, 46)

One major factor that accelerated the growing demand for bone collection was the spread of Social Darwinist theory. When, at the end of the nineteenth century, social Darwinists predicted the inevitable dying-out of "primitive" peoples, the collection of skeletal material assumed an additional importance and urgency. What Patrick Brantlinger refers to as "the funereal but very modern science of anthropology" was particularly insistent in emphasising the necessity for Western science "to learn as much as possible about primitive societies and cultures before they vanish[ed] for ever" (2003, 5). As Dr John Bostock solemnly warned the Melbourne Medical Congress in 1923 "[p]osterity will judge us adversely if we let ... ["the 'Grand Old Man' of the human race"] die without taking his full and complete measure" (Qtd. in Murray 2007, 13).

Offering, as they claimed to, the physical proof of why, as sub-human, evolutionary retards, Indigenous peoples were doomed to disappear, collections of cranial and skeletal remains were an essential component of what is now referred to as "salvage ethnography." The anticipated imminent extinction of native peoples was, on the one hand, then, as Brantlinger asserts "the primary motivation" (2003, 5) behind nineteenth-century enthusiasm for bone-collecting and, on the other, as Coombes suggests, a discursive strategy to legitimise the exhibition of human remains (140).

Edward Ramsay, the curator of the Australian Museum from 1874 to 1894, was a man who played a crucial role in encouraging the perception of Aboriginal bodies as valuable anthropological exhibits (Turnbull 1991, 112). In a pamphlet entitled "Hints for the preservation of specimens of natural history," Ramsay, having explained in some detail how to go about preserving the skulls and brains of Australian mammals, concludes with the suggestion that "[t]he brains of Aborigines so prepared would be of great value" (Qtd in Turnbull 1991, 113). Realising that, given the genocidal effects of white settlement, such desirable items as the "skins, skulls and skeletons of Aborigines, males and females" (113) were likely to be "in dwindling supply" (114) henceforth, the curator, explains Turnbull, used Aboriginal remains as "a unique and persuasive currency, to obtain rare specimens of fauna from other parts of the globe, and also to procure 'specimens' of other 'dying' races" (114).

The value that accrued to the dead bodies of Indigenous people as the numbers of living autochthones dramatically declined in the colonised world led to modes of conduct in the scientific community that can best be described by Thomas Buckley's phrase "shattering demonstrations of the role of indifference to the suffering of the other" (1989, 438). It is that indifference that I want to discuss in the final section of my paper.

Ethical implications

The extinction of Tasmania's native population meant, as Helen MacDonald explains, that they "could now only be known through the physical remains they had left behind" (2005, 83). Thus, when William Lanney died in 1869 his status as Tasmania's last full-blood male Aborigine ensured that his body acquired, post-mortem, a significance that the living man had never possessed. On his death, Lanney's body immediately became the object of fierce rivalry between two competing scientific institutions. William Crowther, honorary medical officer at the General Hospital where the corpse was taken, had made a prior arrangement with London's Royal College of Surgeons to process the skeleton and ship it to them. Realising, however, that the Colonial Secretary was preparing to hand over the coveted remains to Tasmania's Royal Society, Crowther took the step of removing Lanney's skull from his corpse and replacing it with that of another cadaver. That this was not simply the aberrant act of a maverick working to his own fanatical agenda but a practice (albeit clandestine) approved of in scientific circles of the time, may be inferred from a letter written by one of Britain's leading physical anthropologists Joseph Barnard Davis in 1856 in which he explains to his correspondent that he knows how to "abstract skulls from dead bodies without defacing them at all, and could teach any medical gentleman to do this" (Qtd. in MacDonald 2006, 109). As MacDonald comments:

It was clearly a well-honed technique and required private access to at least two dead bodies, one from which to abstract the desired skull and another whose skull could be inserted in its place to disguise the theft. (109)

When the resident surgeon of the hospital, George Stokell, informed the Royal Society of the skull swap, he was instructed by the Society's Honorary Secretary to cut off Lanney's hands and feet in order to prevent Crowther from obtaining a perfect skeleton should he decide to return to the morgue to retrieve the rest of the Aborigine's remains. The final episode in this gothic horror story occurred the morning after Lanney's funeral when it was discovered that his grave had been robbed and the 'foreign' skull discarded. In an account published some time later in the Tasmanian Times denouncing Hobart General Hospital's Board of Management (and Stokell), for the role they had played in the mutilation of Lanney's corpse, Crowther claimed that the Aborigine's exhumed body had been returned to the hospital where all the skin and flesh had been removed from its bones leaving "masses of fat and blood ... all over the floor and upon a large box which had been used as a table" (MacDonald 2006, 150). Dr Crowther's—to say the least—unprincipled behaviour in this sordid affair did not prevent him from becoming the first Australian to be awarded a gold medal and a fellowship of the Royal College of Surgeons. Stokell, for his part, had to content himself with the tobacco pouch he had had made out of Lanney's skin.[37]

The desecration of Lanney's corpse, filled Truganini, the so-called "last remaining female Aborigine," with absolute terror. Desperate that her own body should not become the object of such gruesome abuse, she pleaded (to no avail as it turned out) with a clergyman and a physician not to allow her remains to be dissected for scientific processing and display (Cove 1995, 51). Though, due to the "strong community sentiments" (51) aroused by the Lanney debacle, the government ensured that Truganini's remains initially received a proper Christian burial, the Aboriginal woman's bones were, a mere two years later, disinterred, boiled, temporarily stored in an apple-crate, then later exhibited in a glass case in the Tasmanian Museum where they remained for a century (Brantlinger 2003, 129; Hughes 1996, 424). In 2002 samples of her hair and skin hitherto the property of Oxford's Royal College of Surgeons were returned to Tasmania (*The Age*, n.p.). There are two points I wish to emphasise concerning the cases of Lanney and Truganini. One is the gratuitousness of the abuse to which the bodies of the two Aborigines were subjected. As Brantlinger reminds us, "No advance in scientific knowledge came from [the] grisly fiasco [of William Lanney's dismemberment]" (2003, 129); as for Truganini's remains, they "remained unstudied and unmeasured, until, in the 1890s, a new curator almost threw them away by accident" (129). Indeed, Cove points out, "A review of the literature prior to 1950 shows no evidence that Truganini's skeleton had been used in any published research" (1995, 147).

The second point is that the bodies in question could not possibly have been construed as anonymous or abstract biological material. They were the remains of well-known personalities, individuals who were perfectly familiar to Tasmania's settler community. Seen at the time as the last remaining members of their race, both Lanney and Truganini had become local celebrities, recognisable fixtures on Hobart's waterfront where white visitors asked them to pose for photographs. They spoke some English, wore the European clothes of the period and, in the case of Truganini at least, had apparently been converted to Christianity. The governor of the time, Sir Charles Du Cane recalled Truganini's visits to Government House where she partook of cake and wine and, as the governor facetiously observed, "looked every inch a queen" (Osman 2004, n.p.). "King Billy" as Lanney was mockingly dubbed, had even been introduced to Prince Alfred, the Duke of Edinburgh when HRH visited Hobart in 1868. Most importantly, no-one in Tasmania was oblivious to the fact that these two individuals had witnessed and/or experienced first-hand the forced removals, abductions, murders, rape, "dispersals" and mass die-outs from introduced diseases to which the Indigenous peoples of all white settler colonies were subjected during the genocidal phase of colonisation.

Tasmania's men of science did not live on a different plane of existence, sealed off from the rest of settler society. They were as aware as anyone in the colony that Lanney and Truganini were individuals on whom unspeakable suffering had been inflicted. Moreover, as MacDonald reminds us,

> ...medical men have always understood the bodies upon which they go to work to be ambiguous material: recently subjects, now objects, called 'subjects' for dissection.
> This is because human remains matter. Every society has conventions for dealing with them in a way that involves regulating who has access to bodies and care in their disposal. And when medical men work on the dead, they do so in this knowledge (2006, 3).

Yet, the treatment which the remains of William Lanney and Truganini received at the hands of the scientific community was devoid of any such human understanding. As Tasmania's last "full-blood" natives, the two Aborigines were mere objects of anthropological inquiry, raw data to be processed and, perhaps above all, valuable anatomical trophies. For, like other colonial collectors, William Crowther, explains MacDonald, was involved in a 'culture of exchange' with British men of science like Joseph Barnard Davis and William Henry Flower. In this exchange,

> Flower and Davis received the kind of important objects that boosted their own standing in scientific circles, enabling them to publish on the Tasmanians. In return their colonial collectors gained what they valued most, which was acknowledgement that they were something more than forgotten men living in an outlandish place. (2006, 110)

A parallel example of the "shattering" indifference of the European scientist to Indigenous suffering is that of the story of Ishi, the so-called "last California Indian". As Thomas Buckley suggests, in the eyes of A. L. Kroeber, the anthropologist who both befriended and studied Ishi, after his discovery in Oroville in 1911, the "disease, malnutrition, forced removal, massacre, aggravated rape, flogging and hanging" (1989, 438) that had resulted in the extinction of Ishi's people were mere "small-scale historical experience" that could not be allowed to obscure the "millennial sweeps and grand contours" of social evolutionary time (Kroeber qtd in Buckley 1989, 439).

Consequently, despite the Indian's manifest fear of crowds, Kroeber allowed him to be exhibited in the University of California's anthropology museum where he spent the five remaining years of his life as a living specimen. According to Scheper-Hughes the room Ishi was given "was located next to a hall housing a large collection of human skulls and bones that appalled and depressed [him]" (2002, 358). And later, despite the revulsion his Indian "friend" had expressed at the white man's practice of preserving human remains for scientific investigation, Kroeber, who was in New York when Ishi's body was autopsied after his death from tuberculosis in 1916, did not hesitate to offer the Indian's brain to the National Museum's collection (2002, 361).[38] Yet, as was the case for Lanney's remains half a century earlier, there is no evidence, says Scheper-Hughes "that Ishi's brain was ever included in any physical anthropological or scientific study. It was simply forgotten and abandoned in a Smithsonian warehouse, kept in a vault of formaldehyde with several other brain 'specimens'" (361).[39] In recent years, institutions such as Oxford's Pitt Rivers Museum and the Glasgow Museums have similarly affirmed that during the century in which they held Indigenous remains prior to repatriating them, no request had ever been made to study them (Jones 1994, 29; Lovelace 1994, 30). MacDonald similarly points out that the collection of Indigenous remains possessed by the British Museum (until recently "the most recalcitrant in repatriating Aboriginal skeletal material"), was in the nineteenth century, described as being " 'composed of skulls filthy with dust, and in a dark cellar' " (2006, 184).

How, then, are we to interpret the unflinching callousness that Western science displayed towards the human occupants of the 'new worlds' Europeans invaded during the imperial era? There are, I would like to suggest by way of conclusion, two main alternatives. Either we situate such Olympian detachment within a frame of reference that glorifies knowledge as an absolute value and the pursuit of knowledge as a self-evident good (Cove 1995, 177)—in which case we normalise the indifference of scientists to the extreme psychological suffering of their fellow beings. Or we use a more human yardstick and acknowledge that, even in relation to the period in which it occurred, the behaviour scientists adopted in their study of Indigenous bodies was socially deviant, unethical and profoundly inhumane.

It is important not to lose sight of the fact that both imperialism itself and the human sciences that flourished in the nineteenth and early-twentieth centuries depended on practices that were fundamentally at odds with the moral, philosophical and religious principles that Western societies professed belief in at that time. If European nations were able to justify the crimes against humanity white settlers committed in the colonised territories of empire, it was only because, through a discursive sleight of hand, the indigenous inhabitants of those territories were represented as falling outside the bounds of humanity proper.

By establishing taxonomic schemes classifying the so-called "primitive" races of mankind as fossils, fauna or the missing link, by reducing Indigenous bodies to the status of data, by exhibiting osteological specimens in museum displays that placed Aboriginal peoples at the lowest level of human evolution, and by promoting the knowledge constructed by the human sciences as having the same authority and validity as the "norm of truth" epitomised by Newtonian physics (Poovey 1993, 259), anthropology and its associated disciplines proved themselves the willing helpmates of the imperial enterprise. As to the indifference Victorian men of science displayed in the face of Indigenous suffering, it should be read, I think, not just as the magisterially aloof disinterest from which science necessarily views the objects of its study, but rather as a perversion of the civilised human behaviour that European societies claimed to champion—the consequence, as Buckley would have it, of functioning in an academic economy in which the "'data'[derived from Indigenous bodies] and the taxonomic systematization that they enabled were forms of capital" (1989, 440).

Works Cited

Adas, Michael. 1989. *Machines as the Measure of Men: Science, Technology and Ideologies of Western Dominance.* Ithaca and London: Cornell U. P.

Asad,Talal. 1973. *Anthropology and the Colonial Encounter.* Amherst (NY): Humanity Books.

Brantlinger, Patrick. 2003. *Dark Vanishings: Discourse on the Extinction of Primitive Races, 1800-1930.* Ithaca (N.Y.) & London: Cornell U.P.

Buckley, Thomas. 1989. Suffering in the Cultural Construction of Others: Robert Spott and A.L. Kroeber. *American Indian Quarterly.* Vol. 13, No. 4: 437-445.

Chesterman, John and Brian Galligan. 1997. *Citizens Without Rights: Aborigines and Australian Citizenship.* Cambridge: Cambridge U. P.

Coombes, Annie. 1994. *Reinventing Africa: Museums, Material Culture and Popular Imagination.* New Haven & London: Yale U.P.

Cove, John J. 1995. *What the Bones Say: Tasmanian Aborigines, Science and Domination.* Ottawa: Carleton U.P.

Diamond, Stanley. 2004. Anthropology in Question. *Dialectical Anthropology.* Vol. 28:11-32.

Fabian, Johannes. 1983. *Time and the Other: How Anthropology Makes its Object.* New York: Columbia U.P.

Gould, Stephen Jay. 1981. *The Mismeasure of Man.* New York & London: W. W. Norton & Co.

Harris, Neil. 1995. Museums and Controversy: Some Introductory Reflections. *The Journal of American History.* Vol. 82, No. 3. Dec.: 1102-1110.

Hitchcock, Robert K. 2002. Repatriation, indigenous peoples, and development lessons from Africa, North America, and Australia. *Pula: Botswana Journal of African Studies.* Vol. 16. No.1: 57-66.

Hughes, Robert. 1996. *The Fatal Shore: A History of the Transportation of Convicts to Australia, 1787–1868.* London: Harvill Press.

Jackson John P. Jr. & Nadine M. Weidman. 2004. *Race, Racism and Science: Social Impact and Interaction.* Santa Barbara (Cal.): ABC-CLIO.

Jones, Schuyler. 1994. Crossing Boundaries. *Museum's Journal.* Vol. 94 No. 7. July: 29.

Jopson, Debra and Tony Stephens. 2000. The White Blindfold View. *The Sydney Morning Herald.* 10 June.

Kociumbas, Jan (ed). 1998. *Maps, Dreams, History: Race and Representation in Australia.* Sydney: Braxus Publishing.

—. 2004. Genocide and Modernity in Colonial Australia, 1788-1850. In *Genocide and Settler Society*. Ed. A. Dirk Moses. New York: Berghahn Books, 77-102.

Lorimer, Douglas. 1996. Race, science and culture : historical continuities and discontinuities, 1850-1914. In *The Victorians and Race*. Ed. Shearer West. Aldershot: Scolar Press, 12-33.

Lovelace, Antonia. 1994. A Special Case. *Museum's Journal*. Vol. 94. No.7: 30.

MacDonald, Helen. 2005. Reading the 'Foreign skull': An Episode in Nineteenth Century Colonial Human Dissection. *Australian Historical Studies*. Vol. 125: 81-96.

—. 2006. *Human Remains: Dissection and its Histories*. New Haven & London: Yale U.P.

Morrison, James. 2002. The skeletons of colonialism may get a decent burial at last. *The Independent*. November 10.

Murray, Caitlin. 2007. The 'Colouring of the Psychosis': Interpreting Insanity in the Primitive Mind. *Health and History*. Vol. 9. No 2:1-20. http://www.historycooperative.org/journals/hah/9.2/murray.html Accessed 01/13/2009.

Osman, Andrys. 2004. Truganini's Funeral. *Island*. No. 96, Autumn. http://www.islandmag.com/96/article.html Accessed 01/13/2009.

Palmer, Alison. 2000. *Colonial Genocide*. Adelaide: Crawford House.

Phillips, Ruth B. 2006. Show Times: de-celebrating the Canadian nation, de-colonising the Canadian museum, 1967-92. In *Rethinking Settler Colonialism: History and Memory in Australia, Canada, Aotearoa New Zealand and South Africa*. Ed. Annie E. Coombes. Manchester: Manchester U.P., 121-39.

Poovey, Mary. 1993. Figures of Arithmetic, Figures of Speech: The Discourse of Statistics in the 1830s. *Critical Inquiry*. Vol. 19. No. 2: 256-76.

Reynolds, Henry. 1974-75. Racial Thought in Early Colonial Australia. *The Australian Journal of Politics and History*. Vol. XX. No. 1. April: 45-53.

Rennie, David. 1998. Museums Defended over Bodies Claim. *The Daily Telegraph*. March 19.

Rowley, C.D. 1974. *The Destruction of Aboriginal Society*. Ringwood (Vic.): Penguin Books.

Rydell, Robert W. 1999. 'Darkest Africa' African Shows at America's World's Fairs 1893-1940. In *Africans on Stage: Studies in Ethnological Show Business*. Ed. Bernth Lindfors. Bloomington (Indiana): Indiana U.P., 135-155.

Scheper-Hughes, Nancy. 2001. Ishi's Brain and Ishi's Ashes: Anthropology and Genocide. *Anthropology Today*. Vol. 17. No. 1, February: 12-18.

—. 2002. Coming to our senses: Anthropology and Genocide. In *Annihilating Difference: The Anthropology of Genocide*. Ed. Alexander Laban Hinton. Berkeley: U of California Press, 348-381.

Shelton, Anthony Alan. 2000. Museum Ethnography: An Imperial Science. In *Cultural Encounters: Representing Otherness*. Ed. Elizabeth Hallam & Brian Street. London & NY: Routledge, 156-194.

Simpson, Moira. 1996. *Making Representations: Museums in the Post-Colonial Era*. London: Routledge.

Slogget, Robyn. 2005. 'I have now made a start ....': Dr Leonhard Adam's Ethnographic Collection at the University of Melbourne. *Open Museum Journal*. Vol. 7. November: 1-32.

Southworth, Edmund. 1994. A Special Concern. *Museums Journal*. July: 23-25.

Stander, Jack. 1993. The 'Relevance' of Anthropology to Colonialism and Imperialism. In *The 'Racial' Economy of Science: Toward a Democratic Future*. Ed. Sandra Harding. Bloomington (Ind): Indiana U.P., 408-427.

Stepan, Nancy. 1982. *The Idea of Race in Science: Great Britain 1800 - 1960*. Macmillan Press.

Tatz, Colin. 1999. Genocide in Australia. AIATSIS Discussion Paper, No. 8: 1-50.

The Age. "Editorial," May 30 2002.

Tobias, P.V. 2002. Saartjie Bartman: Her life, her remains and the negotiations for their repatriation from France to South Africa. *South African Journal of Science*. Vol. 98. No.3-4. March- April: 107-110.

Turnbull, Paul. 1997. Ancestors, not Specimens: Reflections on the Controversy over the Remains of Aboriginal People in European Scientific Collections. *The Electronic Journal of Australian and New Zealand History*. April 27.
http://www.jcu.edu.au/aff/history/articles/turnbull.htm
Accessed 01/13/2009.

—. 1991. 'Ramsay's Regime': The Aboriginal Museum and the Procurement of Aboriginal Bodies, c. 1874-1900. *Aboriginal History*. Vol. 15. Nos. 1-2: 108-121.

SCIENCE AND THE AMERICAN EMPIRE: THE AMERICAN SCHOOL OF ANTHROPOLOGY AND THE JUSTIFICATION OF EXPANSIONISM

DONNA SPALDING ANDRÉOLLE AND SUSANNE BERTHIER FOGLAR

The role of science in the founding and expansion of the American empire began with European voyages of discovery in the late 15th century. Science not only led to the classification of "discovered" populations into inferior racial groups, but also served as the measuring stick of civilisation: what was perceived less scientifically or technologically advanced than European cultures was deemed "primitive." American expansionism of the 19th century was no exception to this rule; the rush to lay claim to the vast territories included in the Louisiana Purchase—or to territories that would be the object of annexation later in the century—couched its justification in the deployment of multiple scientific pursuits such as charting topographical features, gathering information on native inhabitants and above all classifying all possible resources necessary to future (White Anglo-Saxon) settlement. The Lewis and Clark Corps of Discovery expedition of 1804-1806, the Zebulon Pike expedition of 1806-1807, or the Wilkes expedition of 1838-1842 clearly illustrate the importance given to such scientific endeavors by the American government of the time.

The object of this paper will thus be to attempt to understand in what ways the scientific approach to race through craniology as practiced by Samuel B. Morton, and later by his followers Josiah Nott and George Gliddon, contributed to the divisive issues of slavery on the one hand and that of Indian removal on the other. And yet polygenism—the theory of multiple origins of man—and the American School of Anthropology can only be understood in the larger context of American cultural attitudes towards race which are deeply rooted in the unique combination of religious worldview and the political pragmatism of cohabitation with the Other. One has but to observe the mixture of Biblical overtones and down-to-earth claims to nationhood in the Declaration of Independence.

Anthropological visions of Otherness will also share this double focus: Morton and his contemporaries were constrained by the Christian dogma of man's creation while being, at the same time, engaged in the practicalities of collecting, classifying and measuring human crania which provided scientific evidence to the contrary. What could have been Morton's claim to fame—making craniology an accepted branch of medical science—was unfortunately "highjacked" by ideologues who used his research to promote their own political agendas.

Religion and Otherness

Manifest Destiny was originally justified by the Scriptures. In Genesis 1:28, Adam and Eve are blessed by God who tells them to "Be fruitful, and multiply, and replenish the earth, and subdue it: and have dominion over the fish of the sea, and over the fowl of the air, and over every living thing that moveth upon the earth" (King James Bible.) This was usually understood by the early colonists as relating to the spreading of White Christian civilisation. America was "discovered" at the same time as the Protestant reformation emerged in Europe; England's Puritans were part of a similar spiritual background, believing England had not broken with the ways of popery (Stephanson 1999, 4). They were imbued with a Biblical vision of their journey across the Atlantic as a reenactment of the Exodus, when the persecuted Jews—God's chosen people of the Old Testament—fled Egypt to find Canaan, going as far as comparing the King of England to the Egyptian Pharaoh, thereby reinventing the Jewish notion of chosenness, migration, and redemption (Stephanson 1999, 10). Several passages of the Old Testament mention the fact that the chosen people were to expand at the expense of the heathen (Stephanson 1999, 25). On the new continent, the Puritans set out, in the words of their leader John Winthrop, to create "a city on the hill," a New Jerusalem, an example for all to see. Chosenness was therefore part of the first utopian colonisation of the (Anglo-Saxon) new world. When they started to establish the New England colonies, the Puritans found evidence of native presence while the indigenous tribes seemed to have largely vanished. In the 1630s, the natives suffered from devastating smallpox epidemics and the ensuing depopulation was for John Winthrop "a miraculous plagey" (Stephanson 1999, 11).

In the 173 years between the founding of the New England colonies and the Louisiana Purchase, the colonies, and later the United States went through a rapid process of territorial expansion. While the first century witnessed a sluggish coastal expansion, the influx of settlers and the prospect of free land gradually drew the immigrants to the west where the King of England tried to stop their expansion at the crest of the Appalachians by setting the limit of colonial settlement at the Proclamation Line in 1763. In 1776, the pressure to expand was such that the western limit became one of the grievances expressed in the Declaration of Independence. At the end of the Revolution, the 13 colonies, now states, were granted territorial extension to the Mississippi; the birth of the new nation with a new and untried political system increased the pace of territorial extension. In 1803, the Louisiana Purchase nearly doubled the size of the nation. It seems therefore inevitable that the nationalist sentiment carried strong overtones of chosenness and that chosenness was inextricably linked to the white race of European origin.

The nation that extended westward with relative ease excluded the colored others in her midst: black slaves and Native Americans, as both were outside the national consensus. According to Senator Benjamin Leigh of Virginia (writing in the Jacksonian era, approximately 1820):

> It is peculiar to the character of this Anglo-Saxon race of men to which we belong, that it has never been contented to live in the same country with any other distinct race, upon terms of equality; it has, invariably, when placed in that situation, proceeded to exterminate or enslave the other race in some form or other, or, failing that, to abandon the country. (Stephanson 1999, 27)

In matters of slavery, expansionist journalist John O'Sullivan believed that

> In respect to the institution of slavery itself, we have not designed, in what has been said above, to express any judgment of its merits or demerits, pro or con. National in its character and aims, this Review abstains from the discussion of a topic pregnant with embarrassment and danger—intricate and double-sided—exciting and embittering—and necessarily excluded from a work circulating equally in the South as in the North. It is unquestionably one of the most difficult of the various social problems which at the present day so deeply agitate the thoughts of the civilized world. Is the negro race, or is it not, of equal attributes and capacities with our own? Can they, on a large scale, co-exist side by side in the same country on a footing of civil and social equality with the white race? In a free competition of labor with the latter, will they or will they not be ground down to a degradation and misery worse than slavery? (O'Sullivan, 1845)

O'Sullivan is also known for having invented the "American multiplication table": [...] "the fulfilment of our manifest destiny to overspread the continent allotted by Providence for the free development of our yearly multiplying millions." (O'Sullivan, 1845). The "yearly multiplying millions" were, of course, Anglo-Saxons.

From sacred to secular: Viewing race in the "empire of liberty"

The founding texts of the American Nation mentioned slaves and Indians as belonging to inferior races or mentioned them as borderline individuals who existed despite the Anglo-Saxon advance. In the Preamble of the Declaration of Independence (1776), the founders of the Nation explained that "we hold these truths to be self-evident, that all men are created equal." And yet, in the list of grievances against the King of England, they accused him of having

> endeavored to bring on the inhabitants of our frontiers, the merciless Indian Savages, whose known rule of warfare, is an undistinguished destruction of all ages, sexes and conditions.

Of course, there remains the question of what people like Jefferson meant by the expression "all men are created equal," in a country where slavery was an established institution. Stanton argues that Jefferson believed in the abolition of slavery and thought that the path to abolition was through science which would demonstrate the equality of men:

> [...] Jefferson remained unshaken in his confident faith that science, by proving men equal, would banish subjection from the world. Regretfully declining an invitation from the citizens of Washington to join the other Signers in attendance at their celebration of the fiftieth anniversary of the Declaration of Independence, Jefferson wrote what was to be his final will and testament to his countrymen. Surveying the contemporary scene, he found everywhere evidences of the beneficent effects of the Declaration. 'All eyes are open or opening to the rights of man,' he noted serenely. 'The general spread of the light of science has already laid open to every view the palpable truth, that the mass of mankind has not been born with saddles on their backs, nor a favored few, booted and spurred, ready to ride them legitimately, by the grace of God.' (Stanton 1960, 23).

In fact, the problem of the "peculiar institution" would not be resolved at the time of the drafting of the Constitution (1789) and black slaves appeared only in population statistics, counting for 3/5ths of a white man, while Indians were counted only if they paid taxes ("the whole number of free persons [...] excluding Indians not taxed, three fifths of all other Persons," Article I, section 2). In the *Northwest Ordinance* (1789), a document written at the same period as the ratification of the Constitution, the question of slavery and the Indian tribes was taken up again, stating this time that "there shall be neither slavery nor involuntary servitude in the said territory" while, paradoxically, it provided for the return of fugitive slaves to their owners. Indians, no longer the "merciless savages" of the Declaration of Independence, had to be taken care of with "the utmost good faith" (Article 3 of the *Northwest Ordinance*).

Thus, the founding documents of the nation illustrate the paradox of racial attitudes, an American vision torn between abstract philosophical considerations of equality and the lived experience of racial cohabitation in the United States. Although the founders were inspired by Locke's assertions of equal status among creatures of the same species and rank without subordination or subjection, William Stanton makes the point that "Locke never lived in proximity to Negroes or Indians. Americans did, and many wondered just how far the concept of equality extended" (1960, 2).

Races and spaces in *Crania Americana* and *Types of Mankind*

Science and medical science

Medical science and science in general, witnessed an upstart in the 19th century. Moving away from the traditional "bedside medicine," the new medicine became centered on laboratory work and statistics as well as on the acquisition of data. This was the so-called Paris medicine, characterized by physical examination, routine autopsy, use of statistics, techniques that could be "pursued properly only in a hospital setting;" in such thought the doctor perceived the patient no longer as a person but as a body (Jacyna 2006, 54-55). When Morton began his work on *Crania,* this particular form of medicine had started to gain acceptance in the United States.

Other factors also impacted Morton's views on racial difference and the scientific methods to be used in determining them. Morton grew up in an era where science and scientific advances were focused on statistical evidence: for example the Lewis and Clark expedition was expected to produce accurate cartography as well as detailed statistical information on the Indian tribes they encountered such as population numbers, body measurements, food habits and life expectancy; Dr. Benjamin Rush's instructions to Lewis included taking the body temperature of tribe members and monitoring their heart rate:

[…] The state of the pulse as to frequency in the morning, at noon & at night—before & after eating? What is its state in childhood, Adult life & old age? The number of strokes counted by the quarter of a minute by glass, and multiplied by four will give its frequency in a minute. (Ronda 2002, 2)

The rise of the so-called "American School" of Anthropology was situated in the context of a fast-expanding American republic, of the growing tensions between North and South over the issue of slavery (connected directly to the integration of western territories as "free" or "slave" states) and the policies of Indian removal from areas east of the Appalachian Mountains. To complicate matters further, American anthropological thought in the 18th century had been concerned simultaneously with disproving European theories of the degenerating impact of the climate in the New World, promoting the notion of man's natural equality despite the "peculiar institution" and seeking a scientific answer to the then confused speculations on the origins of mankind:

At that time the Bible was recognized as the standard by which to arbitrate all moral and ethical judgments, but those with a vested interest in the slave population found biblical arguments to support their own position. The ninth chapter of Genesis relates how the black races were descendants of Ham's son Canaan, who was cursed by Noah, and seems to indicate that they were forever destined to be the white man's servant. […] A less popular school of argument (the polygenists) abandoned the Bible altogether and maintained that the races were separate biological species. […] Morton's single purpose of mind was dedicated to clear the air of this emotional issue and provide hard objective data showing the intelligence of Negro, white man and, for good measure, the North American Indian. (Taylor, online version)

Two of the main debates about race in this early period focused on the blackness of the Negro's skin:

Dr. Benjamin Rush, having determined that the Negro's color was a symptom of endemic leprosy, drew three conclusions. First, whites would cease to 'tyrannise over them,' for their disease 'should entitle them to a double portion of our humanity.' However, by the same token, whites should not intermarry with them, for this would 'tend to infect posterity' with the 'disorder.' Finally, attempts must be made to cure the disease. (Stanton 1960, 12-13)

At the same time others theorized as to how the Indians had arrived in America from the site of the Creation, and as to the origins of the moundbuilders of the Mississippi Valley (of which sample crania were later to figure in Samuel Morton's *Crania Americana*).

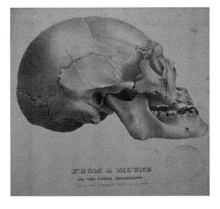

"From a mound on the upper Mississippi" *Crania Americana*, plate 52

One of the "missions" of the western expeditions of the early 19th century was the investigation of the physical bases of racial distinction; Titian Peale and John Kirk Townsend procured Indian skulls and the skeletal remains for collections, curated at the Academy of Natural Science in Philadelphia, by Samuel George Morton (Porter 2003, 70). Morton's scientific reputation had first been made through the publication of *Synopsis of the Organic Remains of the Cretaceous Group of the United States* published in Philadelphia in 1834 in which he described the invertebrate fossils collected during the Lewis and Clark expedition of 1804-1806 (Stanton 1960, 26). As American interest in the subject of race mushroomed in the early 1830s, Morton delivered public lectures on topics ranging from the origin and diversity of the human species to the brain size of various races, racial abilities, the origin of black skin color, cranial characteristics of Negroes and human racial hybrids (Porter 2003, 71).

Morton, from a Quaker background, was a doctor with two medical degrees, one from the University of Pennsylvania Medical School and the other from the University of Edinburg in Scotland (Gillett 1987, 22);[40] he had become interested in crania in the 1820s while preparing a lecture for his anatomy classes because he could find no information on the subject. This led him to establish, over the following years, an impressive collection of more than one thousand complete crania dubbed "America's Golgotha" in scientific circles; and his meticulous study of them was to

result in the publication, in 1839, of his famous work *Crania Americana*.
According to William Stanton in *The Leopard's Spots: Scientific Attitudes toward Race in America 1815-59*:

> Skulls were not an unusual collector's item in the 1830s; [...] the romantic interest they stimulated, their use in medical schools, and the rage for phrenology in the first half of the 19th century, led many people to collect crania. (1960, 29)

Other than Morton's collection, there were two other major collections at the time: one in Cincinnati belonging to Joseph Buchanan; and one in Boston belonging to Dr. John Collins Warren with whom Morton corresponded and exchanged crania. (Stanton 1960, 10)

Morton was not a field naturalist, relying on expedition members and army surgeons stationed in frontier outposts to obtain specimens;[41] he based his evaluations of such crania on these correspondents' descriptions of the circumstances of the discovery of a particular skull, its situation in the earth, the geological conformation of the site, and their opinion as to the tribe to which it belonged. This practice, Stanton notes, which would be "inexcusably lax in a modern anthropologist, was not considered untoward in the early part of the nineteenth century." (1960, 28)

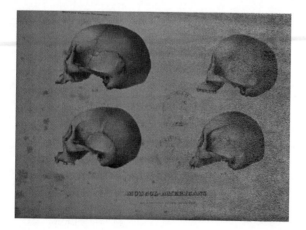

An example of 4 "Mongol American" crania from *Crania Americana*, plate 70

Today there is controversy over how these crania were obtained, as most of the acquisitions were taken from Indian burial sites, some of them from prehistoric mounds, others from recent burial sites, still others harvested on battlefields thanks to his correspondence with army surgeons.

> That army surgeons stationed at remote western outposts and explorers […] took the trouble, often at great hazards to themselves (for some tribes had strong taboos against the desecration of the dead), spoke eloquently of the wide reputation that Morton's collection had acquired." (Stanton 1960, 28)

Morton lamented the fact that some specimens were more difficult to obtain than others. In the preface of *Crania Americana*, he points out that his collection of Mexican crania—and the engravings of them in the book—were not as numerous as he would have liked, due to "the extreme difficulty of obtaining authentic crania of those people" (v).

Craniology as racial theory

Craniology is the scientific measurement of cranial capacity and conformation; measurement, as part of the new statistical medicine, appeared as the impartial way of establishing a nomenclature of racial characteristics. Morton's method was both empirical and comparative: he made 13 different measurements of each cranium in his possession, including the interior capacity for which he practiced a painstaking method using white pepper seed, described at length in *Crania Americana*; he then compared the results based on Johann Blumenbach's five-race theory of anthropology, showing in Table I page 290 that the Caucasian race had the largest mean internal capacity, followed by the Mongolian race, the Malay race, the American race (i.e. Indians) and lastly the Ethiopian race (i.e. the Negro). As scientific and objective as these measurements may have appeared at first glance, in fact they were 'skewed' to fit pre-existing racial views: for example, Morton did not factor in gender and body size, and intentionally excluded the study of Hindu skulls in the Caucasian category because it lowered the average cranial capacity. He also used a disproportionally high number of Peruvian skulls in the "American race" category even though these particular crania were the smallest in the group (Menand 2001, 110).

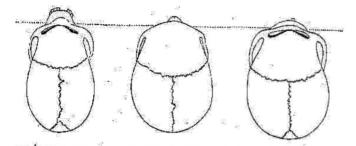

The *first* of these figures represents a Negro head, elongated, and narrow in front, with expanded zygomatic arches, projecting cheek bones, and protruded upper jaw. The *second* is a Caucasian skull, in which those parts are nearly concealed in the more symmetrical outline of the whole head, and especially by the full development of the frontal region. The *third* figure is taken from a Mongol head, in which the orbits and cheek bones are exposed, as in the Negro, and the zygomæ arched and expanded; but the forehead is much broader, the face more retracted, and the whole cranium larger. Having been at much pains to give the *norma verticalis* of the skulls figured in this work, the reader will have ample opportunity to compare for himself. He will see that the American head approaches nearest to the Mongol, yet is not so long, is narrower in front, with a more prominent face and much more contracted zygomæ.

Sketch of Negro, Caucasian and Mongol crania with racial observations by Morton

Although Morton specifically stated that he was not competent in the field of phrenology, a task left to George Combe in the appendix of *Crania Americana* entitled *Phrenological Remarks on the Relation Between the Natural Talents and Dispositions of Nations, and the Development of their Brains*, this did not prevent him from making sweeping statements on the racial characteristics of American Indians and Blacks in his work that would be fully exploited after his death by Josiah Nott, George Gliddon and Louis Agassiz. For example, he is famous for declaring, in the Introductory essay, that "in their mental character the Americans are averse to cultivation and slow in acquiring knowledge; restless, revengeful and fond of war, and wholly destitute of maritime adventure" (Morton 1839, 6); or again in the subchapter dedicated to the conclusions drawn from measuring the crania of the "Appalachian branch" of Indians (i.e. all those of North America excluding the Mexicans) finding that "in character these nations are warlike, cruel and unforgiving. They turn with aversion from the restraints of civilized life, and have made but trifling progress in mental culture or the useful arts" (64). The idea of converting tribes to civilisation through the practice of agriculture dated back at least to Thomas Jefferson; declaring them racially inept to do so, combined with the racial propensity for war if one is to believe Morton, surely lent a 'scientific' justification to Indian removal policies. He thus concludes his chapter on the American Race with the conviction that

> One of the most remarkable intellectual defects of the Indians is a great difficulty in comprehending anything that belongs to numerical relations. [...] Mr. Schoolcraft, the United States Indian Agent, assures me that this deficiency is a cause of most of the misunderstanding in respect to treaties entered into between our government and the native tribes. The latter sell their land for a sum of money without having any conception of the amount, so that if it be a thousand dollars or a million few of them comprehend the difference until the treaty is signed and the money comes to be divided. Each man is then for the first time acquainted with his own interest in the transaction, and disappointment and murmurs invariably ensue. (Morton 1839, 83)

Yet it must be noted as a final point that Morton warned against the dangers of seeking to enslave Native Americans: "but it must be borne in mind that the Indian is incapable of servitude, and that his spirit is sunk at once in captivity, and with it his physical energy [...]" (Morton 1839, 75).

The same cannot be said about Morton's evaluations of the "Negro Race." Besides the fact that the Ethiopian Race constitutes the lowest category in Morton's five-race classification mentioned earlier, the introductory remarks on the various African groups inform the reader that "the moral and intellectual character of the Africans is widely different in different nations" (87). This is followed by a list of racial characteristics ranging from "intelligent and industrious" to "remarkably stupid and slothful." Despite differences observed in specific tribes, Morton makes the generalizing statement that "the Negroes are proverbially fond of their amusement, in which they engage with great exuberance of spirit; and a day of toil is with them no bar to a night of revelry;" or that "they appear to be fond of warlike enterprises, and are not deficient in personal courage; but, once overcome, they yield to their destiny, and accommodate themselves with amazing facility to every change of circumstance" (87).

The Negro race, therefore, was scientifically proven to be the perfect one for enslavement since, in opposition to the Indian, the Negro was "pliant, capable of yielding to his fate, accommodating himself to his condition, [and bearing] his heavy burthen [sic] with comparative ease" (75). These observations could not go unnoticed in the decade preceding the Civil War; *Crania Americana* opened the door for phrenologists to associate cranial shape with brain size, and brain size with mental capacity and social station (Porter 2003, 70), inadvertently providing fuel for the pro-slavery politics of the South.

From craniology to phrenology and beyond

What Morton had begun, others would expand using his work; in 1854, Josiah Nott and George Gliddon published *Types of Mankind, Ethnological Researches based upon the Ancient Monuments, Paintings, Sculptures and Crania of Races and upon their Natural, Geographical, Philological and Biblical History*, in fact a study of the races based not only on *Crania Americana* but also on Morton's second, smaller volume *Crania Ægyptiaca* (1844), a study of a collection of crania from Egyptian tombs which had been provided to Morton by George Gliddon. The most notable contributions of this more mature work of research were the discovery of the great age of races which surpassed the theories then in vogue and Morton's declaration that slavery "was among the earliest social institutions of Egypt" (59, in Stanton 1960, 51).

Josiah Nott, a medical doctor from Mobile, Alabama and ardent admirer of Morton's work, took it upon himself after Morton's death in 1851 to develop the theory of polygenism already hinted at, but not confirmed outright, in *Crania Americana*. *Types of Mankind*, based on the statues, frescos and remains found in Egyptian tombs, as well as on papers contributed to the authors by the late Samuel Morton's wife, attacked theological theories of the unitary origins of man and rejected the Biblical chronology of man's existence. The work, in fact, contained little new information, as it was designed as a compendium of all anthropological evidence that had been brought to light in support of the specific diversity of mankind (Stanton 1960, 163). It came as no surprise that Nott in particular, a pro-slavery Southerner, was seeking proof of the Negro's inferiority and racial predestination as slave labor, echoing Morton's conclusions both in *Crania Americana* and in *Crania Ægyptiaca*. According to Nott,

> For the sake of illustrating that, even in Ancient Egypt, African slavery was not altogether unmitigated by moments of congenial enjoyment; not always inseparable from the lash and the hand-cuff; we submit a copy of some Negroes 'dancing in the streets of Thebes' (Fig. 185), by way of archaeological evidence that, 3400 years ago (or before the Exodus of Israel B.C. 1322) 'de same ole Nigger' of our southern plantations could spend his Nilotic Sabbaths in salutary recreations [...] (Nott et.al. 1855, 263)

This statement is accompanied by an illustration of Negro slaves dancing in an Egyptian fresco which occupies the right-hand side of the page 263 and the quote "Turn about, and wheel about, and jump Jim Crow!" (see below)

Nott also shared Morton's view that the impact of white civilisation on the Negro and the American Indian was totally different:

> while the contact of the white man seems fatal to the Red American, whose tribes fade away before the onward march of the frontier-man like the snow in spring (threatening ultimate extinction), the Negro thrives under the shadow of his white master, falls readily into the position assigned him, and exists and multiplies in increased physical well-being. (Morton 1839, xxxiii)

Morton, however, had pointed out that "a system of encroachment and oppression has been practiced upon them since the first landing of Europeans on the shores of America" (78) and had seemed genuinely concerned by the probable extinction of the American race. Nott, less than 20 years later, relates his personal experience of the Indians in his home town in far less philanthropic terms:

> We see every day, in the suburbs of Mobile, and wandering through our streets, the remnant of the Choctawa race, covered with nothing but blankets, and living in bark tents, scarcely a degree advanced above brutes of the field, quietly abiding their time. No human ingenuity can induce them to become educated, or to do an honest day's work: they are supported entirely by begging, besides a little traffic of the squaws in wood. To one who has lived among American Indians, it is in vain to talk of civilizing them. You might as well attempt to change the nature of the buffalo. (1855, 69-70)

In a political context in which expansion of the American empire necessarily implied government intervention in solving the "Indian problem," the writing of Nott, Gliddon and later Louis Agassiz (also a contributor to *Types of Mankind*) come as no surprise. At the same time, their repeated attacks on Biblical interpretations on the origins of man, so essential to the social fabric of the South, weakened their contributions to the American School of Anthropology, to such an extent that Southerners were willing to reject their scientific justifications of slavery. In Stanton's words,

> the career of the American School was an ironical one, for despite their incisive criticism of theories of development they helped to prepare the public mind for the Darwinian chronology. [..] The final irony was that Darwinism took on added luster for having supplanted the 'infidel' doctrine of the American School. (1960, 196)

Morton's lack of interest in the political implications of his work and the presence of Combe's phrenological interpretations of his collection in the appendix of *Crania Americana* seem to have condemned him somehow to scientific oblivion, if one is to point out that his two works have never been reedited since their original publications in 1839 and 1844 respectively. After Morton's death in 1851, Nott and Gliddon's intensive political use of Morton's work to defend slavery in the final years before the Civil War as well as in the era of spectacular Westward expansion contributed significantly to giving Morton a bad name in the field of anthropology. Surprisingly enough, according to Vincent Sarich and Frank Miele, authors of *Race: The Reality of Human Differences*,

> The most extensive analysis of the American School of Anthropology has shown that it actually had little effect on nineteenth century America's most divisive issue. For the North and for all opponents of slavery, the issue was one of morality, not science. Nor did the South embrace polygenism as a scientific defense for keeping blacks in servitude. Southerners relied instead upon tradition and religion (2004, 76-77).

It was thus that the American School of Anthropology, begun by Morton and brought full circle by Nott, Gliddon and Agassiz, would end its claim to posterity as a footnote to Darwin's *The Descent of Man* (1871).

Acknowledgements

The authors of this article wish to thank the Bibliothèque de Médecine de l'Université Joseph Fourier Grenoble I for granting access to their copy of *Crania Americana*

Bibliography

Primary sources

King James Bible. online version, http://etext.virginia.edu/kjv.browse.html>, accessed March 17, 2009.

Morton, Samuel George. 1853. Catalogue of Skulls of Man and the Inferior Animals Etc. *The Southern Quarterly Review* 8: 59-92. Making of America.

—. 1844. *Crania Aegyptiaca. Observation from Aegyptian Ethnography Derived from Anatomy, History, and the Monuments.* Philadelphia: Penington.

—. 1839. *Crania Americana; or, a Comparative View of the Skulls of Various Aboriginal Nations of North and South America. To Which Is Prefixed an Essay on the Varieties of the Human Species.* 1st ed. Philadelphia: J. Dobson.

Nott, Josiah. 1855. *Types of Mankind or Ethnological Researches Based Upon the Ancient Monuments, Paintings, Sculptures, and Crania of Races, and Upon Their Natural Geographical, Philological, and Biblical History.* Philadelphia: Lippincott, Grambo, and Co.

O'Sullivan, John L. 1845. Annexation. *United States Magazine and Democratic Review.* 17. July–August: 5-10.

Winthrop, John. "A Model of Christian Charity." 1630. lib.virginia.edu. accessed March 17 2009.
<http://religiousfreedom.lib.virginia.edu/sacred/charity.html>.

Secondary sources

Gillett, Mary C. 1987. *The Army Medical Department, 1818-1865.* Washington D.C.: Center of Military History, U.S. Army.

Jacyna, Stephen. 2006. Medicine in Transformation 1800-1849. In *The Western Medical Tradition.* Eds. W. F. Bynum, et al. New York (N.Y.): Cambridge University Press: 11-101.

Menand, Louis. 2001. Morton, Agassiz, and the Origins of Scientific Racism in the United States. *The Journal of Blacks in Higher Education.* Online version accessed March 17 2009.

Porter, Roy. 2003. *Blood and Guts: A Short History of Medicine.* New York: Norton.

Ronda, James P. 2002. *Lewis and Clark among the Indians.* Lincoln, NE: University of Nebraska Press.

Sarich, Vincent, and Frank Miele. 2004. *Race: The Reality of Human Differences.* Boulder: Westview Press.

Stanton, William. 1960. *The Leopard's Spots - Scientific Attitudes toward Race in America 1815-59.* Chicago: The University of Chicago Press.

Stephanson, Anders. 1999. *Manifest Destiny. American Expansion and the Empire of Right.* New York: Hill and Wang.

Taylor, Ian. 2003. *In the Minds of Men: Darwin and the New World Order.* Toronto: TFE Publishing. (5th edition). Online edition accessed November 2, 2008.

PART III:

SCIENCE AND POLITICAL DISCOURSE

A "REPUBLIC OF SCIENCE"?
THE ROLE OF SCIENCE IN THOMAS
JEFFERSON'S IMPERIAL VISION

MEHDI ACHOUCHE

Among the Founding Fathers, none is more studied, commented on and even, in some cases, venerated than Thomas Jefferson, author of the Declaration of Independence and third president of the United States (1801-1809). Jefferson concentrates in his person and his own life story what are often seen as the very characteristics of early nineteenth century America: a vibrant apologist of individual liberties and of democracy, Jefferson was on the other hand a slave-owner; a great admirer of Native American cultures, he also was the architect of their extermination and one of the masterminds behind the western expansion of the nation. More than anything else, Jefferson was a statesman haunted by the thirst for knowledge, impassioned with the sciences of his time, the friend of the French Ideologues and one of the most eloquent apologists of the Enlightenment. Nevertheless this scientific curiosity and activism has very often been treated lightly by historians, who tend to give them a marginal role compared to Jefferson's more grandiose political undertakings, having them illustrate his enlightened genius and mindset but not bear directly on U.S. history. Yet science was more than an enlightened pastime: it lay at the heart of Jefferson's political projects and of his vision of an imperial America. Jefferson brought science to the fore at a critical time in American history, a time when the current ideals and the values of the nation crystallized.

This was also the time when the "Republic of Science" declared its independence: after having progressively separated itself from the Republic of Letters in the course of the 18th century, it too founded itself on ideals of equality, fraternity and universality between men of science across the world. It also rested on a sophisticated epistolary network connecting individuals, learned circles and societies separated by political and geographic borders, on the steady publication of learned societies' transactions and proceedings, and on the publication of the journals of the expeditions that were so popular at the time. The apologist of the young

American republic, Jefferson also was an ardent citizen of the new republic of science, the untiring literary correspondent of countless scientists across the world, a member of the most important learned societies on both sides of the Atlantic, and was himself the author of notable works in the field of archeology and paleontology. Not himself a scientist, he was nonetheless a great promoter of science, as one of his biographers has written (Bedini 1990, 491).

Jefferson being thus at the crossroads between science and politics, the notion of a republic of science and its potential polysemy offers a glimpse into Jefferson's thoughts and achievements as they pertain to the relationship between these two worlds. The republican notion stresses Jefferson's essentially agrarian definition of a perfect republic, and the way he characteristically tried to reconcile this Arcadian vision with his scientific and progressive undertakings. It also allows for the confrontation between the philanthropic and cosmopolitan ideal conveyed by the notion of a republic with a more nationalist, perhaps even imperialistic, vision of scientific knowledge.

The Arcadian Republic

Thomas Jefferson occupies an important place in the history of ideas for his unflinching apology of agrarian values and his vision of a predominantly Arcadian America, a bucolic ideal which lies at the heart of Jeffersonian republicanism.[42] Chapter XIX of Jefferson's *Notes on the State of Virginia*, written in the early 1780's, thus figures today as the axiomatic expression of American agrarianism, reflecting the standpoint of many at the time and denying the need for manufactures in the young republic:

> Those who labour in the earth are the chosen people of God, if ever he had a chosen people, whose breasts he has made his particular deposit for substantial and genuine virtue. It is the focus in which he keeps alive that sacred fire, which otherwise might escape from the face of the earth. (Jefferson 1984, 290)

Or the following excerpt from his correspondence:

> Cultivators of the earth are the most valuable citizens. They are the most vigorous, the most independent, the most virtuous, & they are tied to their country & wedded to its liberty & interests by the most lasting bonds. As long therefore as they can find employment in this line, I would not convert them into mariners, artisans or anything else. (Jefferson 1984, 818)

Similar sentiments can be found in the writings of many other intellectuals of the era and found their most prominent expression in the celebrated *Letters from an American farmer*, by John Hector de Crèvecoeur (1782). Benjamin Franklin himself, perhaps the greatest American scientist of the times, subscribed to this vision and declared with satisfaction: "The great Business of the Continent is Agriculture. For one Artisan, or Merchant, I suppose, we have at least 100 Farmers, by far the greatest part Cultivators of their own fertile lands" (Smith 1950, 141). These declarations are well-known, as well as Jefferson's attachment to the agrarian world, connecting democracy, republicanism and the agricultural world. Yet their implications for the status of the sciences in America are less frequently examined. Jefferson of course does not repudiate science when he lauds the rural virtues, but he does express an extreme defiance towards science's practical applications in the burgeoning industries, in so far as they are thought to threaten the virtue underpinning the American republic, a virtue, as the previous quotation makes clear, only to be found in the countryside. This idea had been familiar since Antiquity and was famously advocated in the seventeenth century by James Harrington in his *Commonwealth of Oceana* (1656). Jefferson, a Virginian and owner of several plantations, deeply committed to the interests of the agrarian South, was relentless in his attempts to put science to the service of agriculture, studying and lobbying on behalf of horticulture, chemistry, mineralogy or climatology, trying every time to use his influence and access to the international scientific community to the service of national agriculture. Jefferson thus worked all his life for the setting up of agricultural societies throughout the colonies, designed to help stimulate farming innovations and to better spread information within the federation; corresponded tirelessly with people across the world to exchange information and seeds; and introduced on the American continent numerous new plants and animal species with high economic potential (Bedini 1990, 393, 493). He even invented a new type of mouldboard that won him a gold medal from the Société d'Agriculture de Paris in 1806.

For his political opponents, themselves fierce supporters of the new industries, Jefferson's agrarianism amounted to conservatism and betrayed his "visionary" character.[43] Yet Jefferson never lived this parallel attachment to a rural socio-economic model and to scientific progress as a paradox, entertaining a conciliatory vision and never explicitly connecting the manufactures he criticized so much in his correspondence with his advocacy of scientific progress. When confronted with the political necessity of industrialization, Jefferson dreamed up an ideal representation of manufactures implanted on farms and in the countryside, a domestic and rural industrialization which would blend naturally with the landscape and espouse agrarian values, the "machine in the garden" as Leo Marx famously characterized the dream in his book. Jefferson could all the more deny the backwardness of his ideas as he was far from being the only one advocating agriculture as an economic asset: the French Physiocrats, some of whom were friends and correspondents of Jefferson (most markedly Du Pont de Nemours) advocated reliance on agriculture rather than commerce, the champion of traditional mercantilist theories.[44] Adam Smith himself, in his celebrated *Inquiry into the Nature and Causes of the Wealth of Nations* (1776), read by Jefferson, enjoins Americans, in the name of the international division of labor, not to import manufactures and to specialize in agriculture. Yet Jefferson was not an economist, and his attachment to the agrarian world is first and foremost moral and emotional, the countryside shunning "the Cloacina of all the depravities of human nature" to be found in cities and to create a new Arcadia harmoniously uniting nature and the concept so dear to the Enlightenment, civilization.[45] Such a vision can be found in the quasi-utopian project of Jefferson to build new cities, or expand existing ones such as New Orleans, on a scientific, checkerboard plan, every white square of the board being devoted to public squares and greens. The project was set up as a means to avoid new yellow fever epidemics, but it also betrays Jefferson's desire to find a physical and symbolic compromise between urban and rural spaces (White 1977, 17), and foreshadows the urban plans of late nineteenth-century American utopians.[46] Denying any causal relationship between life conditions in the great urban centers he abhorred so much and the ongoing industrial and technological revolution, Jefferson wanted to believe that the American republic would be able to avoid the British fate, which he had witnessed while in Europe in the late 1780's, and would reconcile technological progress and agrarian life. The secret of this American exceptionalism lay, in Jefferson's mind, in the West, in the infinity of arable lands waiting for the American farmer to deliver their cornucopian wealth and allowing the agrarian republic to clone itself endlessly, that is, without having to endure any fundamental change in its socio-economic structure, and without going through the urban development so harmful to Europe. The thesis that was to become gospel in the American mind and cause so much anxiety to Frederick Jackson Turner once the colonization of the West was thought to be over, in 1893.[47]

Once the industrialization of the country was on its way, movements advocating a return to nature and to a pre-industrial state were to multiply in America, often taking the form of rural communities of varying inspirations. It is interesting to make a parallel between these and Jeffersonian thought: all have in common their rejection of urban life, and if few of them explicitly reject the machine age, still all advocate a return to a mode of life hardly compatible with the industrialization of the country, a fact giving to all of them (Fourier's phalanges included) an anti-urban, anti-industrial, hue.[48] And even though Jefferson, faced with Napoleonic wars and their impact on the country, evolved with the times and ultimately had to admit the necessity for the country to industrialize so as to become economically independent, he never gave up on his belief that rural life was morally best and that somehow, America would be able to juxtapose what for some were two distinct phases of one linear process of economic evolution (including Adam Smith). Jefferson himself endorsed the developmental theory late in his life and applied it to the United States, describing to a correspondent the spatial as well as chronological evolution separating the "savages" of the western part of the continent from the metropolis of the eastern coast, the latter being described as the pinnacles of civilization, in fundamental contradiction with his repeated denunciations of the decadent urban world and thus betraying an ambiguous vision of civilization's fruits (Jefferson 1984, 1497).[49] Jefferson for most of his life denied this historical evolution, thinking science as only confirming and reinforcing America in its natural state, lauding the progress of reason but refusing to contemplate how the machines he saw in action in England were destined to revolutionize the nature of his country. It is not only hindsight which gives the modern reader this advantage over Jefferson; for all his attachment to scientific advance, Jefferson, as other intellectuals of his time, clung to a static vision of society and expressed the almost instinctive revulsion of many late-eighteenth, early nineteenth century writers when confronted with the dawn of the Mechanical Age. Jefferson's words and agrarian dream would long influence American thought and would long be used to justify both an optimistic faith in progress and the necessity to reinstate the country's rural values. It is this fascination for Jefferson's irenic vision, giving the nation an ideal representation of itself so openly in contradiction with actual facts, giving the country the prospect of a final, post-historical reconciliation between progress and tradition, which entitles the Jeffersonian vision to be called a national utopia, setting forth the enticing prospect of an Arcadian scientific republic.

The Empire of Science

Another tension characterizes Jefferson's understanding of scientific progress. A fierce advocate of cooperation between scientists across the world, subscribing to Condorcet's famous project of a "universal republic of science" (1988, 303), based on full equality and transnational solidarity of the scientific community, Jefferson was nevertheless a consummate nationalist in the scientific domain as in all others. For the declared architect of a future American empire, the country's future greatness necessarily entailed scientific greatness.

It is not so much the practical value of the sciences that was so important for Jefferson, although it did have an important role to play, as the international prestige already pertaining to scientific discoveries at the time.[50] The notion of civilization included scientific advances, and every 'polished' nation had to put forward its own achievements in the domain. The phenomenon acquired added significance in the 1740's, a decade that saw the founding of a great many learned societies in Europe and which saw, in 1743, the foundation by Benjamin Franklin in Philadelphia of the American Philosophical Society (APS), which was presided by Jefferson from 1796 to 1814, at the same time that he was vice-president and then president of the country. It is precisely the epistolary network connecting these societies and academies throughout the world, a great number of which made Jefferson an honorary member, which constituted for Jefferson the Republic of Science:

> These societies are always at peace, however these nations be at war. Like the republic of letters, they form a great fraternity spreading over the whole earth, and their correspondence is never interrupted by any civilized nation. (Jefferson 1984, 1201)

Once in the White House, Jefferson would use the APS to launch a great number of expeditions of discovery, using those sponsored by England's Royal Society as a model, most prominently Cook's circumnavigations. Destined to explore the *terra incognita* of the trans-Mississippi West, to sound out the economic potentialities of the region and to contact local Native American tribes, thus fulfilling eminently practical uses, these expeditions, including the most famous of them, the Lewis and Clark Expedition, were also destined to fulfill a more symbolical role, marking America's place in the ranks of polished nations that were actively and spectacularly contributing to the advancement of knowledge and to the spread of civilization worldwide. Reproducing, sometimes even mimicking, those earlier explorers' actions was part of the ritual of exploration and discovery for American explorers, who, on Jefferson's insistence, duly noted and published the accounts of their journeys and discoveries, signifying to the world the level of scientific sophistication reached by the American nation. This transparency of information desired by Jefferson, which contrasted so markedly with the Spanish authorities' great secrecy when it came to geographic information, must thus not be considered purely from a philanthropic perspective. This symbiosis between political, military and scientific objectives was common at the time, finding its most accomplished fulfillment in Bonaparte's Egyptian expedition, even though it was not about geographic exploration. As René Sigrist, in a recent semantic analysis of the terms "Republic of Science" and "Empire of Science," writes: "Compared with a Republic of Letters that aimed to be universal and disinterested, the Empire of science thus appears as an object of power and rivalries between nations," the notion of empire, at first only a synonym for domain or area, slowly transforming into a political notion (Sigrist 2008, 346). The sciences contribute to empire building, not only in their practical and measurable contribution, but also in the prestige pertaining to the nation that will best be able to don civilization's mantel. And who better than the United States, in direct contact with the "savages," to take over the civilizing mission.

But American identity then defined itself not only via its supposed superiority over indigenous tribes, but also with respect to the nations of the Old World, where much criticism was then formulated on the supposed inferiority of Americans themselves. Mid-way as the American nation was located, in-between western savages and eastern enlightened nations, the question for some was whether Americans were not themselves the barbarians. Worse, Comte de Buffon, the famed French naturalist, as well as the Abbé Raynal and several other European writers, basing themselves on environmental theories positing a possible influence of the natural environment (climate, vegetables, nature of the soil, etc.) on the development and form of biological productions, raised the issue of the possible degeneracy, physical but also possibly moral for some writers, of American biological productions, including human beings. This famous transatlantic debate between European advocates of and American opponents to the theory used the whole scientific arsenal of the time to buttress the arguments of both sides, from paleontology to the various natural sciences. Jefferson used his presence in Paris as the American ambassador to France (1785-1789) to meet Comte de Buffon and try to convince him of his error, notably by having the carcass and horns of an American moose sent from America to Paris to convince Buffon of his mistake (Bedini 1990, 148-151). Science was thus used as the instrument for a partially ideological debate, the motivations of some being sometimes at least as much emotional and ideological as scientific and factual. Behind the controversy can be detected, before and after the American revolution, an implicit debate on the cooptation of civilization, some Americans (and their European friends) being close to thinking that civilization had departed Europe and settled on the new continent. Condorcet could thus declare, after the start of the French revolution, that civilization had reached its highest point in France and the United States, the new step in humanity's progress being defined by the advent of liberty and democracy (Condorcet 1988, 266). While Americans could exhibit their technological sophistication to western tribes, they could also, increasingly, represent themselves as the chronologic as well as spatial vanguard of civilization.

The only element wanting for this representation to be working was to prove with facts, to Europeans as well as, maybe, themselves, their civilized greatness, and above all other things their scientific greatness. To the Abbé Raynal, stressing the fact that America had thus far produced no man of genius, Jefferson, an untiring apologist of the American cause, made this famous answer in his *Notes on the State of Virginia*:

> In war we have produced a Washington [...] In physics we have produced a Franklin, than whom no one of the present age has made more important discoveries, nor has enriched philosophy with more, or more ingenious solutions of the phaenomena of nature. We have supposed Mr. Rittenhouse second to no astronomer living: that in genius he must be the first, because he is self-taught. As an artist he has exhibited as great a proof of mechanical genius as the world has ever produced. He has not indeed made a world; but he has by imitation approached nearer its Maker than any man who has lived from the creation to this day. (Jefferson 1984, 190-191)

David Rittenhouse was a Philadelphia astronomer and clockmaker whom Jefferson ceaselessly promoted overseas as the quintessential example of a sublime American scientific success. Rittenhouse's masterwork, of which Jefferson is speaking in the above quotation, was a mechanical planetarium, the miniature representation, not only of the solar system, but also of the movement of the planets for any point in time within a timeframe of 5,000 years. The perfect representation of a universe imagined as a mechanical device, as was the custom of the day, the precision of these planetariums symbolized the exactness with which God had made the world, but also the exactness of Enlightenment science, which explains their immense popularity at the time. The individual, and by extension the nation, capable to best reproduce the work of the "great clockmaker" would thus be endowed with the prestige that only expeditions of discovery were then susceptible to produce. Jefferson thus asked Rittenhouse to fabricate a new planetarium, with the idea of sending it to Louis XVI in order to prove to Europe the greatness of American science, a project which never materialized (Hindle 1974, 170-171, 338-339). Jefferson, great promoter of his country that he was, never missed an opportunity to prove that American science was that of a great country, destined to become an "Empire for Liberty," an expression he liked to use (Jefferson 1903, 277).

This last expression, which betrays Jefferson's wish to see America become a great nation while remaining true to the cause of liberty, is all the more interesting as it points to a correlation, in the mind of a great many Americans at the time as well as Jefferson's, between scientific greatness and freedom. The future American greatness, the future American empire, was indeed all the more required as the country had been founded on the promise of a radiant future, the American nation showing the way to the world, the familiar idea of American exceptionalism. Mentioned previously was thus Condorcet's endorsement that the advent of democracy marked a higher level of progress. However, the idea of an imperial destiny for the country, taking over from older empires, had been in the air since at least the eighteenth century and gained impetus with each new victory of the Americans over European foes, first during the French and Indian War (1754-1763) against the French and then against the British in the War of Independence. In a world dominated by tyranny and superstition, America, the eldest daughter of the Enlightenment whose *raison d'être* was liberty and reason, had to shine in the political and scientific domains, becoming an empire struggling for the spread of liberty and scientific truth, thus proving the validity of the democratic experience. Jefferson could thus write to the president of Harvard University, thinking of the next generation of Americans:

> We have spent the prime of our lives in procuring them the precious blessing of liberty. Let them spend theirs in shewing that it is the great parent of *science* and of virtue; and that a nation will be great in both, always in proportion as it is free. (Jefferson 1984, 949, original italics)

America, according to poet Joel Barlow in 1787, was "the empire of reason. Here neither the pageantry of courts nor the glooms of superstition have dazzled or beclouded the mind" (quoted by Hindle 1974, 253).

The correlation between liberty, reason and science can be found everywhere in the texts of the era, even before independence. Philip

Freneau and Hugh Brackenridge, in their ode to "the rising Grandeur of America," thus described America as "a land of ev'ry joyous sound / of liberty and life; sweet liberty! / Without whose aid the noblest genius fails / And science irretrievably must die." (quoted by Hindle 1974, 249). Jefferson in his correspondence thus tirelessly correlated the advent of democracy in America with scientific progress, both being signs of the perfectibility of mankind, "freedom & science" being the two greatest achievements of the revolution (Jefferson 1984, 1066), freedom itself being characterized by him as "the first-born daughter of science" (1023). William Dunbar, a correspondent of Jefferson and himself to join one of his later expeditions, thus hoped that "Arts, Sciences, and Literature may take a flight, which will at length carry them as far beyond those of our European brethren, as we soar above them in the enjoyment of national liberty" (quoted by Petersen 1970, 763). For Jefferson as for many others, science could not develop fully in Europe because of the despotisms sapping the continent and because of papal censorship, forcing scientists such as Joseph Priestley to seek refuge in America, "just as the pope imagined when he shut up Galileo in prison that he had compelled the world to stand still" (quoted by Petersen 1970, 579). A 1780 sermon could thus note that the arts and sciences "delight in liberty, they are particularly friendly to free States," while another apologist noticed, in 1785, that:

> The introduction and progress of *freedom* have generally attended the introduction and progress of *letters* and *science*. In despotick governments the people are mostly illiterate, rude, and uncivilized; but in states where CIVIL LIBERTY hath been cherished, the human mind hath generally proceeded in improvement, – learning and knowledge have prevailed, and the arts and sciences have flourished. (John Gardiner, An Oration delivered Jul 4, 1785, *Boston Gazette*, May 9, 1785; quoted by Hindle 1974, 251)

Welcoming a reason which had had to exile itself from Europe, America, this nation "advancing rapidly to destinies beyond the reach of mortal eye," as declared Jefferson in his first inaugural address, was thus destined for a scientific greatness that would validate the efforts of the revolutionary generation (Jefferson 1984, 492). Hence the need, felt by the American scientific community throughout the nineteenth century, to impose itself on the international stage, not only to prove itself, but also to equal in power and prestige European colonial empires and demonstrate America's exceptional greatness. Science and scientific preeminence, as much in practice as symbolically, were then part and parcel of the imperial project.

I began by stressing the fact that Jefferson was often representative of the aspirations, the values and the paradoxes of the American nation as a whole, and I tried to show that the same held true for his relationship to science. The "apostle of Americanism" as one historian has written, Jefferson symbolizes and illustrates the idiosyncratic welcome given to the scientific enterprise in America for at least 200 years.[51] If science did fulfill a practical role in Jefferson's vision, it is first as a mode of representation that science is most relevant to a study of the beginnings of the American nation, as one of the main arguments in a discourse specific to the United States and to the first of its ideologues. The empire, in its territorial definition, was already forming, and science did play a key role in the settling of the West. But it is also the empire of the mind that Americans were aiming for, and Jefferson can help answering the question as to why Americans so often think of themselves as the Republic of science *par excellence*.

Works Cited

Bedini, Silvio A. 1990. *Thomas Jefferson: Statesman of Science*. New York: Macmillan Publishing Company.

Condorcet, Jean-Antoine-Nicolas Caritat. 1988. *Esquisse d'un Tableau historique des progrès de l'esprit humain, suivi de Fragment sur l'Atlantide* [1795]. Paris: Flammarion.

Hindle, Brooke. 1974. *The Pursuit of Science in Revolutionary America, 1735-1789*. New York: Norton & Company.

Jefferson, Thomas. 1984. *Writings*. New York: Literary Classics of the United States.

—. 1903. *The Writings of Thomas Jefferson, volume 12*.Washington: The Thomas Jefferson Memorial Foundation of the United States.

Noyes, John Humphrey. 1966. *Strange Cults and Utopias of 19th-Century America (Formerly titled: History of American Socialisms)* [1870]. New York: Dover.

Peterson, Merrill D. 1970. *Thomas Jefferson and The New Nation–A Biography*. New York: Oxford University Press.

Schmitt, Peter J. 1990. *Back to Nature–The Arcadian Myth in Urban America*. Baltimore: The Johns Hopkins University Press.

Sigrist, René. 2008. "La République des Sciences" : essai d'analyse sémantique. In *La République des Sciences. Réseaux des correspondances, des académies et des livres scientifiques* (333-359). Dix-huitième siècle, 40.

Smith, Henry Nash. 1950. *Virgin Land–The American West as Symbol and Myth*. New York Vintage Books.

White, Morton and Lucia. 1977. *The Intellectual Versus The City–From Thomas Jefferson to Frank Lloyd Wright*. New York: Oxford University Press.

SCIENTIFIC RHETORIC AND THE AMERICAN EMPIRE: TWO VERSIONS

JEAN-MARIE RUIZ

In America as in Europe, "science" and "empire" have often been considered compatible, if not interrelated and mutually supporting. But in the United States, former colonies born in the peculiar period of the Enlightenment and determined to build an "empire of reason," their interplay has been more problematic. Due to the United States' initial rejection of classical imperialism, their compatibility was not taken for granted and they had to be reconciled with the dominant values of the New World. This was made possible by the early belief that the *American empire* was altogether different from European colonialism, because it relied on consent and attraction rather than on coercion in a supposedly empty environment. It is by no accident that western expansion was conceived, rationalized and initiated by Thomas Jefferson, one of the foremost representatives of both the Enlightenment and American science. By the time the Pacific was reached however, dominant values had changed beyond recognition and during the next decades, science was increasingly used to legitimize overseas expansion, in spite of its association with traditional imperialism. This paper is an attempt to examine such a sudden and surprising embrace of colonialism through the prism of the scientific—viewed from today's standard, rather pseudoscientific—rhetoric used by American advocates of empire in order to emphasize the difference with the earlier form of American expansionism. As such, it may be seen as a modest contribution to the contemporary scholarly debate on whether continuity between continental and overseas expansion exists[52] (Vincent 1999, 76-81). The analysis of the interaction of scientific rhetoric with broader intellectual trends and the comparison of the different imperial projects they have underpinned will show, I hope, that very similar language and principles may lead to very different conclusions and outcomes. While the influence of science on politics may be viewed as enduring and perhaps unavoidable, much of its impact depends on its association with dominant values, which to a great extent determine the nature of empire.

Science, Empire and the Enlightenment

As was emphasized by historians in the 1960's and 70's, the 18th century was "the age of science" or "the Age of Reason" to borrow the title of Thomas Paine's celebrated book. "Everywhere the scientists were philosophers, and most of the philosophers were scientists" (Commager 1977, 1). At least two Founding Fathers, Franklin and Jefferson, were prominent scientists, for whom politics could also be scientifically practiced. Believing in God as a Master Clockmaker, they thought that both the natural and the social spheres, the Earth not less than the whole universe, were ruled by universal laws that could be discovered through reason and whose knowledge was essential to an enlightened policy:

> How exact and regular is everything in the natural world. How wisely contriv'd... All the heavenly Bodies, the Stars and the Planets, are regulated with the utmost wisdom ! And can we suppose less care to be taken in the order of the Moral than in the Natural System? [53]

Similar reasoning accounts for the idea that "politics may be reduced to a science" to borrow David Hume's famous phrase[54] which, according to Douglas Adair, greatly influenced James Madison, the author of the 10th Federalist paper and Father of the federal Constitution (Adair 1974, 93-106). More generally, the Founders relied on the Scottish Enlightenment anthropology, i.e. on the idea that human nature was universal and constant—therefore predictable—and could be scientifically studied through history (Adair 1974, 95-96). The *Federalist Papers*, were informed by a veritable theory of human nature that substituted self-interest for the classical virtue as the cornerstone of society and politics (Ruiz 2001, 7-16; Scanlan 1951, 657-677; Wright 1949, 1-31).

As Felix Gilbert has shown, the belief that the same rules governed the physical and social worlds also accounts for the golden age of diplomacy in 18th century Europe and its adoption in America once the early radical dream of ending it had waned (Gilbert 1961, 44-137). When independence was declared, many Americans believed with Paine and the radical Whigs that traditional diplomacy was a remnant of the doomed old order, that the crumbling of monarchies and the creation of republics in their stead would substitute commerce for "political connections." The business of republics was not foreign policy, let alone empire building, but the primacy of domestic issues. As Thomas Pinckney said at the Philadelphia convention

> conquest or superiority among other powers is not or ought not ever to be the object of republican systems. If they are sufficiently active and energetic to rescue us from contempt and preserve our domestic happiness and security, it is all we can expect from them. (Farrand 1937, I-402)

Such an outlook was short-lived, however, as the traditional French-American alliance of 1778 shows. John Adams, the very author of the Model Treaty of 1776 (the symbol of the American dream of a "commercial diplomacy") was no idealist and his aim in avoiding political connection was to make sure America would not become the victims of French imperialism as they had been the victim of British imperialism (Graebner 1985, xxv; McDougall 1997, 25). The subsequent American adoption of European diplomatic methods and the influence of the literature of the "interests of the states" on the Founders and on Washington's Farewell Address in particular, is a good illustration of the influence of physics on early American theory of foreign policy (Gilbert 1961, 89-136). In the Age of Reason and Diplomacy, many Americans, including the sanguine proponents of a "new diplomacy" took it for granted that the relations between states were somehow similar to those of the planets. As Thomas Paine wrote in *Common Sense*

> In no instance has nature made the satellite larger than its primary planet, and as England and America, with respect to each other, reverse the common order of nature, it is evident they belong to different systems—England to Europe, America to itself. (1966, 75-76)

Paine's notion of an American "system" reflects the influence of science, but contrary to others who would later use it to advocate expansion, his aim in 1776 was only to prove that American independence from Britain was *natural*, rational, and thus legitimate. Neither Paine nor any other political thinker or practitioner of his generation questioned Pinckney's assertion that "conquest" ought not to be sought by the new independent republics, yet few Americans questioned the right of western settlers to move to and hold territories that they considered unoccupied. And few Americans denied their right to secure their borders against Indians or against European empires bordering their territory. If for no other reasons, extending the American empire was therefore deemed necessary to the safety of the new federal republic, and to ensure the survival of republicanism in a predominantly monarchical world. The very notion of an "empty" North America made expansion both possible and necessary, for other powers coveted western territories that seemed easy to grasp. Expansion could thus be associated with security and republicanism, rather than conquest, and as such made compatible with the values and principles of the Enlightenment, particularly in its Scottish version. For Montesquieu—the icon of the French version—thought expansion would destroy republicanism whereas Hume considered it could survive in a large and expanding territory (Adair 1974, 97-98).

These thoughts were important in the development of Thomas Jefferson's "empire of liberty," the twin goals of which were to secure western expansion—partly for security reasons (Tucker and Hendrickson 1990, 161-162)—and legitimize it by differentiating the American guise of empire from the European one. Jefferson's dilemma was to reconcile his unparalleled territorial ambitions with his distrust of power and his contempt for the traditional forms of empire. As Robert Tucker and David Hendrickson emphasized in their *Empire of Liberty*, Jefferson not only wanted to "conquer without war" but to conquer without power (1990, 20-21; 162), at least without what is now called "hard power," which according to him and the Anti-federalists had a baneful influence on societies. Contrary to Hamilton who shared Frederick the Great's maxim that diplomacy without military power was like music without instruments, Jefferson believed that, in spite of its weakness, the United States could achieve much through diplomacy because the European powers needed American commerce and friendship. The Louisiana purchase proved Jefferson right, and showed the world that the American federal Republic could have both empire and liberty. In accordance with Jefferson's Northwest Ordinances, expanding the Republic would never be tantamount to annexing colonies, and the population of the newly acquired territories would be considered citizens rather than subjects. State equality within the federal Union was to replace the traditional imperial distinction between the metropolis and its colonies (Boyd 1968, 191).

But what about the role of science in Jefferson's expansionist theory and practice? As an icon of the American Enlightenment, Jefferson's links with science are too numerous and obvious to require emphasis. What must be noted here is his use of science to conduct and legitimize the United States' exploration of North America, which was the first step towards western expansion. Just as Great Britain argued that it was "actuated merely by desiring to know as much as possible with regard to the planet we inhabit," Jefferson availed himself of the scientific spirit of the Enlightenment to organize the Lewis and Clark expedition in territories that did not yet belong to the United States but were considered crucial to the future of the Federal Republic[55] (Taylor 2005; Nacouzi 2005). What must also be emphasized is the crucial role of science in legitimizing the American version of empire during the Jeffersonian era and beyond. Since coercion was ruled out, the idea that expansion was a natural phenomenon that, as such, could be scientifically explained, was extremely tempting. Jefferson did not use the notion of political gravitation, but he referred to geography, demography, natural boundaries and natural rights as powerful determinants that one day, would result in "the cover(ing) of the whole northern, if not the southern continent, with a people speaking the same language, governed in similar forms, and by similar laws."[56] Jefferson also contributed to the theory of an American "system" which Paine had already hinted at in the 1770's: "The European nations" wrote Jefferson to Alexander von Humboldt,

> Constitute a separate division of the globe; their localities make them part of a distinct system; they have a set of interests of their own in which it is our business never to engage ourselves. America has a hemisphere to itself. It must have a separate system of interest which must not be subordinated to those of Europe.[57]

These quotes suggest that Jefferson agreed with the substance of Thomas Pownall's assertion twenty years earlier that America had become "a new Primary Planet, which [...] must shift the centre of gravity," even though the influence of Newton's physics in his rhetoric was less obvious (Weinberg 1963, 228). By 1819, John Quincy Adams could also describe the United States' expansion as a natural, almost mechanistic and apolitical phenomenon based on physics:

> [...] the world shall be familiarized with the idea of considering our proper dominion to be the continent of North America. From the time when we became an independent people it was as much a law of nature that this should become our pretension as that the Mississippi should flow to the sea. Spain had possessions upon our southern and Great Britain upon our northern border. It was impossible that centuries should elapse without finding them annexed to the United States; not that any spirit of encroachment or ambition on our part renders it necessary, but because it is a physical, moral, and political absurdity that such fragments of territory, with sovereigns at fifteen hundred miles beyond sea, worthless and burdensome to their owners, should exist permanently contiguous to a great, powerful, enterprising, and rapidly-growing nation. (Charles Francis Adams 1874-77, 4: 438)

At about the same time, according to Albert Weinberg, John Quincy Adams referred to a "law of nature" by which the European colonies in North America were bound to join the United States, and which he subsequently called "political gravitation" in a despatch to the United States' minister to Spain concerning Cuba:

> [...] there are laws of political as well as of physical gravitation; and if an apple severed by the tempest from its native tree cannot choose but fall to the ground, Cuba, forcibly disjoined from its own unnatural connection with Spain, and incapable of self-support, can gravitate only toward the North American Union, which by the same law of nature, cannot cast her off from its bosom. (Weinberg 1963, 228-229)

For those who believed in political gravitation, it was perhaps easy to think that islands, as small political bodies, were bound to gravitate around larger continental political bodies. In the following decades, many expansionists would apply it to Cuba.[58] But John Quincy Adams's phrase was also subsequently construed as a more general "natural law of attraction which makes large bodies overcome smaller ones,"[59] therefore applicable to all territories, continental or overseas, adjacent or not, for "the attraction of a larger and stronger country may prove more potent than that of a nearer but smaller land" (Weinberg 1963, 237).

In any case, as long as Jefferson's principles and the main tenets of the Enlightenment prevailed, there were limits to what the rhetoric of political gravitation could advocate in terms of policy. John Quincy Adams himself discarded expansion as soon as it became clear that it would extend slavery, and would therefore threaten the Union and liberty (Tucker and Hendrickson 1990, 162). In the 1840's, the expansionist ideology of manifest destiny still took seriously Jefferson's rule that "conquest [is] not in our principles" (Weinberg 1963, 283). Indeed, as Thomas Hietala emphasized, the new expansionists "sharply repudiated those who doubted the exceptionalism of the American empire" whose superiority "rested on consent rather than coercion" (191-92). Like Jefferson, they contended that the United States could expand without the traditional imperial burdens of colonialism and militarism. In his writings, O'Sullivan assured his readers that the U.S. empire would never be similar to the Roman or British empires because it would rest on equal states and free people (Hietala 1985,192; Merk 1995, 256-57). Texas was the perfect illustration of what O'Sullivan and others had in mind: enterprising Anglo-Saxons had settled in a supposedly vacant territory, fought for their rights to be independent, and were admitted to the Union as an equal state. To be sure, the acquisition of New Mexico and California did not really follow the same pattern, and the federal government and the military, rather than the pioneers, played the major role. Yet, Albert Gallatin, Jefferson's Secretary of the Treasury and close friend who lived long enough to witness the Mexican war, mainly concentrated his criticism on *All Mexico* advocates who wanted to conquer the entire Mexican territory (Vincent 1999, 51). As a matter of fact, Jefferson would probably have agreed with the strategic reason behind "Mr Polk's war," i.e. conquer the Pacific coast before the British do—even though he would probably have opposed the means and favoured diplomacy and money to acquire the coveted territory, as Polk himself initially tried to do (Hietala 1985, 154-55).

Of course, expansionism of the 1840's did differ from the earlier version, while at the same time remaining faithful to its main tenets, best summarized in Jefferson's "empire of liberty." The most conspicuous change came with nationalism and exceptionalism. Jefferson was nationalist in his own way, and he believed that the United States was destined to rise, but as an American representative of the Enlightenment he was also a citizen of the world whose exceptionalism was rooted in the peculiarity of the American environment rather than the intrinsic virtues or ethnic origins of its dwellers. In contrast, as is well known, the expansionists of the 1840's posited the superiority of the Anglo-Saxons, whose advance they deemed irresistible due to their vigour and virtue, as well as the will of God. What did not change but nevertheless reinforced the new conception was the original myth, strengthened by the Lewis and Clark expedition, "of an empty continent that had been dormant for centuries, awaiting for the energizing presence of Anglo-Americans" (Hietala 1985, 192). While the promoters of a manifest destiny grafted these new trends onto the Founders' expansionist mould, they had also eroded the latter's emphasis on natural rights and equality which shaped the Jeffersonian vision of empire and made consent a prerequisite to expansion—theoretically at least.[60]

Towards 1898

As long as these traditional values informed expansionism, neither overseas acquisitions nor classical imperialism stood any chance of being accepted by Americans or their representatives. Hence the failure of the early proponents of an overseas empire, in spite of their efforts to borrow the scientific rhetoric of their elders. President Johnson and his Secretary of State William Seward referred to John Quincy Adams's law of political gravitation to advocate the annexation of the Danish West Indian islands of St Thomas and St John: "I agree with our early statesmen that the West

Indies naturally gravitate to, and may be expected ultimately to be absorbed by, the Continental states, including our own," said Johnson to persuade Congress to ratify the treaty of purchase of 1867 (Weinberg 1963, 234). Then, in 1870, it was president Grant's turn to ask Congress to annex Santo Domingo, another West Indian Island. But both requests were rejected by Congress, where overseas expansionists were still a minority.

It is only when the new intellectual trends had overwhelmed traditional ones that overseas expansion was accepted and legitimized by yet another kind of scientific rhetoric. In the period during which Mark Twain and Charles Dudley Warner satirically characterized as the *Gilded Age* in their celebrated novel, traditional American ideas about man and society were discarded and science itself played a key role in the process. Even more than during the Enlightenment, faith in science gripped post-bellum America, making it more secular and more materialistic than ever, and substituting a dynamic view of the world and the cosmos in perpetual evolution for their previous conception of a static, orderly and stable universe (Boller 1970, viii). By the 1890's, the departure from traditional values and the loss of old certainties were great enough to cause some thinkers of American extraction—notably Henry Adams—to experience estrangement and anxiety (Parrington 1930, 3: 214-227).

The new age of science was dominated by physics and biology, which had an overall intellectual impact on the social sciences. The discovery of the atom and the cell led to a new perception of the nature of matter that could be used to study the properties of both this world and the cosmos that the telescopes had unveiled. Hence the renewed credit of the Newtonian demonstration that the immeasurable universe was governed by natural laws similar to those governing the earth—such as gravitation— and the renewed temptation to apply these laws to human societies and international relations. Just as scientists assumed that the discovery of atoms and the law of the conservation of energy would allow them to understand the nature of cosmic energy, interpreting the course of human actions and history in terms of universal energy seemed possible, particularly if this was consistent with Darwin's evolutionary theory which, by the 1880's, had won general acceptance on both sides of the Atlantic. This is what the great British philosopher Herbert Spencer set out to do in his seminal work published in 1864, significantly entitled *Synthetic Philosophy*, in which he united the latest findings in biology and physics in a coherent synthesis based on the evolutionary theory (Hofstadter 1959, 31-50). Indeed, as Henry Adams testified in his classic if unusual autobiography, synthesis was a key word among scientists and intellectuals:

The atomic theory; the correlation and conservation of energy; the mechanical theory of the universe; the kinetic theory of gases, and Darwin's Law of Natural Selection, were examples of what a young man had to take on trust [...] The ideas were new and seemed to lead somewhere—to some great generalization which would finish one's clamor to be educated. [...] For the young men whose lives were cast in the generation between 1867 and 1900, Law should be Evolution from lower to higher, aggregation of the atom in the mass, concentration of multiplicity in unity, compulsion of anarchy in order. (Samuels 1973, 224; 232)

Testimonies of the impact of natural sciences, of the trend towards a global synthesis of knowledge, abound in the American political literature of the last three decades of the nineteenth century:

The inspiration of seeing the old isolated mists dissolve and reveal the convergence of all branches of knowledge is something that can hardly be known to the men of a later generation, inheritors of what this age has won

wrote John Fiske (Hofstadter 1959, 12). One of Fiske's friend, Henry Holt, provided a more detailed and telling description of the intellectual environment of his time:

The conception of one power behind all had been a dream of not a few philosophers and poets, but as a fact comprehensible by the average mind, it was not known until the discovery about 1860 of the Conservation of Force. About the same time, was discovered the unity of all organic life, in its descent from protoplasm, and the identity of its forces with those of the inorganic universe. The nebular cosmogony, the persistence of force and the biologic genesis, united together, showed the power of evolving, sustaining and carrying on the entire universe known to us, to be one, and constantly acting in one unified process. (Parrington 1930, 3:207)

For all the intellectual excitement it produced, the scientific synthesis led to conclusions that were running counter to what had been the United States' most cherished principles since its foundation. Darwinism suggested—and social Darwinism proclaimed—that what was bad for the individuals could be good for the race; for Spencer's most influential disciple in the United States, William Graham Sumner, natural rights contradicted the Law of evolution and ought therefore to be rejected. Without inequality, the survival of the fittest was impossible and the only alternative to it was the survival of the unfittest, that is "the law of anti-civilization" (Sumner 1934, 2: 56). According to Sumner, rights were civil, not natural, including individual liberty:

[man has] no more rights than a rattlesnake; he has no more right to liberty than any wild beast; his right to the pursuit of happiness is nothing but a license to maintain the struggle for existence if he can find within himself the power with which to do it. (Boller 1970, 59)

For Brooks Adams, who also used physics and biology to explain universal history and the fate of nations, both individuals and societies were ruled by "forces which [overrode] the volition of man" (1900, 25). The former "must rise or fall in the social scale, according as their nervous system is well or ill adapted to the conditions to which they are born" (1943, 5) whereas the fittest in international relations were those whose energy and concentration gave them a comparative advantage:

From the retail store to the empire, success in modern life lies in concentration. The active and economical organisms survive: the slow and costly perish. Just as the working of this law has produced, during the last century, unprecedented accumulations of capital controlled by single minds, so it has produced political agglomerations such as Germany, the British empire, and the United States. (Brooks Adams 1900, 21-22)

Brooks Adams's theory is "based upon the accepted scientific principle that the law of force and energy is of universal application in nature, and that animal life is one of the outlets through which solar energy is dissipated" (1943, 5-6). According to him, physics and biology made the domination of men or nations over others unavoidable and expansion (or as he calls it, "movement") was natural to a nation and "proportionate to its energy and mass":

The acceleration of movement, which is thus concentrating the strong, is so rapidly crushing the weak, that the moment seems at hand when two great competing systems will be left pitted against each other, and the struggle for survival will begin. (Brooks Adams 1900, 22)

Not all American social Darwinists were expansionists however, and some of the most prominent, like William Graham Sumner, were outspoken anti-imperialists staunchly opposed to overseas expansion. Typically, those on the imperialist side in the 1898 debate combined not only physics and biology, but German political philosophy as well to make their point. As Richard Hofstadter has shown, the United States of the Gilded age was *the* Darwinian country (Hofstadter 1959, 4), but in addition to the influence of British theories, the American post-bellum intellectual landscape was also being transformed by German political philosophy. Vehicles of German influence in America were then numerous and powerful: the prestige of Bismarck's achievements; the hiring of German scholars and the adoption of German academic standards by the most prestigious American universities; the renown of Hermann von Helmholtz, one of the fathers of thermodynamics who had played a crucial role in formulating the famed law of the conservation of energy; more generally the influence of German philosophers and historians such as Hegel, Fichte, Treitschke, Ranke and Mommsen; anglo-saxonism and the idea that the United States and Germany belonged to the same teutonic race; and probably even more important, the similarity between the recent history of Germany and the United States, which had both experienced a strong sense of nationalism.

At about the same time as Social Darwinists, but with even more emphasis, German philosophers and the theorists of *Machtpolitik* promoted the idea that competition, conflict and war were necessary to progress.[61] Their glorification of the nation state entailed a radical critique of natural rights and a related claim that superior peoples were destined to rule those who were either not ready or not capable to self-govern. Among their American disciples, John Burgess did much to undermine the traditional principles dating back to the Enlightenment.[62] Natural rights, he claimed, were "anti-or extra-state rights" which should be discarded as "erroneous and harmful" (Robson 1930, 318). To him liberty was the "creation of the state," whose sovereignty over "individual subject and all associations of subjects" was "original, absolute, unlimited, universal" (Robson 1930, 318; 324). The state's ultimate end was no less than the "perfection of humanity; the civilization of the world; the perfect development of human reason, and its attainment to universal command over individualism; the apotheosis of man" (Robson 1930, 326). His writings show how easily German *Machtpolitik* and social Darwinism— then the most important intellectual trends in the United States—could converge on international relations and imperialism on the eve of the Spanish-American war:

> We advance politically, as well as individually, by contact, competition, and antagonism. The universal empire suppresses all this in its universal reign of peace, which means, in the long run, stagnation and despotism (Robson 1930, 331).
> [...] It is therefore not to be assumed that every nation must become a state. The political subjection or attachment of unpolitical nations to those possessing political endowment appears, if we may judge from history, to be as truly a part of the world's civilization as is the national organization of states. I do not think that Asia and Africa can ever receive political organization in any other way. [...] The national state is... the most modern and complete solution of the whole problem of political

organization which the world has yet produced; and the fact that it is the creation of Teutonic political genius stamps the Teutonic nations as the political nations *par excellence*, and authorizes them, in the economy of the world, to assume the leadership in the establishment and administration of states. [...] The Teutonic nations can never regard the exercise of political power as a right of man. With them this power must be based upon capacity to discharge political duty, and they themselves are the best organs which have as yet appeared to determine when and where this capacity exists. (Hofstadter 1959, 175)

Once the most cogent ideological obstacle to overseas empire, self-determination, had been discarded as irrational and contrary to progress by modern science and political philosophy, the road to 1898 lay open. Those American theorists who could best intertwine the dominant ideological and scientific threads gave continentalism its *coup de grace*, and among them, Germanophiles figured prominently. German political philosophy, particularly *Machtpolitik*, played an important part because the glorification of power and war it provided was necessary to the 1898 watershed but was at odds with the American tradition. Militarists such as Stephen Luce who equated wars with life and peace with death were a tiny minority, and the more numerous advocates of preparedness referred to the old maxim, "if you wish for peace, prepare for war," rather than extol the virtues of war (Hofstadter 1959, 184). Most American writers on war agreed with Spencer that war had been useful in the past or for primitive societies but was no longer useful to modern, advanced civilizations. Those who wished the United States to join the international struggle for empire needed to convince their fellow citizens that it was either necessary to progress, as the German thinkers claimed, or required by forces beyond human control, and therefore by the natural laws that science had revealed.

Both arguments are conspicuous in the writings of Henry Powers, a professor of economics who had lived in Germany and was, of all scholars, the most committed to an overseas empire. According to Powers, George Washington had understood that Americans possessed a "redundant vitality which made them fundamentally and irrepressibly aggressive" and his most important contribution had been to point to "the West rather than the East as the direction in which this surplus of energy could safely spend itself." During most of the 19th century, the United States' behaviour had been "much like that of the boy whose appetite is temporarily satisfied while he was masticating a very large mouthful" (Powers 1898, 176). But once continental expansion was over and the surplus energy had accumulated, traditional obstacles to overseas empire such as isolationism and continentalism were doomed. "That the ideal of national isolation is a Utopia is due to no accident of mood or circumstances, but to laws as fundamental as the constitution of protoplasm" wrote Powers (190), who viewed himself as a "scientist [who], as such has no right to be sorry or glad of anything. His business is to observe phenomena and to study cause and effect" (175). Among these phenomena, was the "fundamental and universal fact, the fact of growth," the "necessary consequence of life" without which, life could not possibly persist (181). Growth however, and therefore life, "means conflict sooner or later":

> The growing aggregates eventually touch, then crowd, and the strong displaces the weak. The incidents of the struggle change, its essence never. Treaties of peace may rule out slugging, but they never stop the struggle. All means are used, all advantages count. By subtle encroachments or violent shock strength displaces weakness without itself knowing why. It may be long before the widening boundaries touch, before the pressure becomes conscious, but the time comes. And then despite all accidents and all precautions, the higher vitality triumphs. It may be a sad fact, but there is no means known by which weakness and inefficiency can inherit the earth. This may or may not be congenial to our moral sense. I have no comment to make on the ethics of expansion. (Powers 1898, 181-82)

While these thoughts reminiscent of Treitschke's *Machtpolitik* were certainly not congenial to the moral sense of many Americans, they were not restricted to marginal thinkers either. Celebrated theorists and historians such as Alfred Mahan, the author of *The Influence of Sea Power upon History*, who exerted considerable influence on President Theodore Roosevelt and other prominent politicians, also wrote on the "ethics of expansion" and used science to legitimize overseas expansion and war. The latter, Mahan acknowledged in an article entitled "The practical aspects of war," was "accompanied by an immense waste of energy and on substance." Yet, he added:

> So is steam; yet just now it is the great motor of the world. Economize, doubtless, to the utmost, by bettering your processes. Reduce the frequency of actual war by such measures as may be practicable; but simultaneously and correlatively make it more efficient, and therefore less wasteful of time and of energy. (Mahan 1907, 166)

While Mahan mainly relied on rational, strategic arguments in his plea to develop the United States naval power and acquire the naval bases necessary to it, he also resorted to science in addition to Malthusian, Darwinian and Hegelian theories of international relations (Ruiz 1999). To be sure, geography was for him one of the main determinants of the policy of nations, but when it came to "the onward movement of the world" (…) both in "rate and in direction," he also relied on physics and biology. His interpretation of history was too dynamic to be based solely on geopolitics. As Brooks Adams or Henry Powers, Mahan viewed the interstate rivalry and the fierce international competition for markets that characterized his time as the result of the long period of European peace and prosperity, of the subsequent increase of its population and of the accumulation of energy it produced (Mahan 1912, 115). Even though he did not believe, as John Hay, that human energy had cosmic origins, his conception of international relations is very similar to Brooks Adams's: we can find in his writings the same image of nations as extending spheres whose speed is determined by their weight and their energy, all converging on areas like China, where the resistance is the least, until they collide (Mahan 1902, 228). Mahan also posited that the proximity of states necessarily involved conflicts, whose outcome ultimately depended on the belligerents' mass and energy (1912, 9).

As for Mahan's solution to the international predicaments of his time, it appears more traditional, but it also relied on physics and is therefore consistent with his analysis of their causes:

> History has shown that in all (…) advance there is an inevitable element of aggressiveness, which can be kept within bounds only by an opposition of force. Thus is ensured a balance, an equilibrium, the maintenance of which has been, and continues to be, the anxious preoccupation of European statesmen. (Mahan 1915, 166)

If balance of power was a traditional solution to what he called "modern conditions," what must be emphasized is that Mahan's use of a scientific rhetoric served the dominant values of his time. Conversely, these stemmed in part from a world view that was shaped by new scientific discoveries that made the political principles of the Enlightenment obsolete. When the United States was born, Newton's physics described an ordered, rational and stable universe that gave rise to renewed faith in the human capacity to pursue happiness and achieve equality. A century later, this was replaced by a world view characterized by constant movement, transformation and universal competition, while biology seemed to prove that progress resulted from inequality and strife. It was amid such an intellectual transformation, to a great extent based on science, that the United States adopted imperialism in its classical guise.

Works Cited

Adair, Douglas. 1974. That Politics May Be Reduced to a Science. David Hume, James Madison and the 10th Federalist. In *Fame and the Founding Fathers*. New York: Norton and Co.

Adams, Brooks. 1900. *America's Economic Supremacy*. New York: MacMillan.

—. 1943. *The Law of Civilization and Decay*. New York: Vintage Books.

Adams, Charles Francis. 1874-77. *Memoirs of John Quincy Adams*. 12 vols, Philadelphia: J.B. Lippincott.

Aron, Raymond. 1984. *Paix et guerre entre les nations*. Paris: Calman-Lévy.

Boller, Paul. 1970. *American Thought in Transition: The Impact of Evolutionary Naturalism, 1865-1900*. Chicago: Rand McNally.

Boyd, Julian. 1968. Thomas Jefferson's "Empire of Liberty". In Peterson, Merrill. *Thomas Jefferson*. New York: Hill and Wang.

Commager, Henry Steele. 1977. *The Empire of Reason*. New York: Oxford University Press.

Farrand, Max. 1937. *The Records of the Federal Convention of 1787*. New Haven: Yale University Press.

Gilbert, Felix. 1961. *To the Farewell Address. Ideas of Early American Foreign Policy*. Princeton, New Jersey: Princeton University Press.

Graebner, Norman. 1985. *Foundations of American Foreign Policy*. Wilmington, Delaware: Scholarly Resources Inc..

Hietala, Thomas. 1985. *Manifest Design: Anxious Aggrandizement in Late Jacksonian America*. Ithaca: Cornell University Press.

Hofstadter, Richard. 1959. *Social Darwinism in American Thought*. Boston: The Beacon Press.

Hume, David. 1953. *Political Essays*. New York: Charles Hendel.

LaFeber, Walter. 1963. *The New Empire: An Interpretation of American Expansion, 1860-1898*. Ithaca: Cornell University Press.

Mahan, Alfred. 1897. *The Influence of Sea Power upon History*. Boston: Little, Brown and Co.

—. 1902. *Retrospect and Prospect*. Boston: Little, Brown and Co.

—. 1907. *Some Neglected Aspects of War*. Boston: Little, Brown and Co.

—. 1915. *The Interest of America in International Conditions*. Boston: Little, Brown and Co.

—. 1912. *Armaments and Arbitration or the Place of Force in the International Relations of States*. New York: Harper and Brothers.

McDougall, Walter. 1997. *Promised Land, Crusader State. The American Encounter with the World Since 1776*. New York: Houghton Mifflin.

Meinecke, Friedrich. 1997. *Machiavellism, the Doctrine of Raison d'Etat and Its Place in Modern History.* Philadelphia: Transaction Books.

Merk, Frederick. 1995. *Manifest Destiny and Mission in American History.* Cambridge, Mass.: Harvard University Press.

Nacouzi, Salwa. 2005. Thomas Jefferson et les raisons d'une expédition : exploration, expansion, expansionnisme. In Caron, Nathalie & Naomie Wulf. *The Lewis and Clark Expedition.* Paris: Editions du temps.

Paine, Thomas. 1966. *Common Sense and Other Political Writings* [1776]. New York: Pyramid Books.

Parrington, Vernon Louis. 1930. *Main Currents in American Thought.* New York: Harcourt, Brace & World.

Powers, Henry. September 1898. The War as a Suggestion of Manifest Destiny. In *Annals of the American Academy of Political and Social Sciences.*

Robson, Charles. 1930. *The Influence of German Thought on Political Theory in the United States in the Nineteenth Century.* Ph.D. History department, University of North Carolina.

Ruiz, Jean-Marie. 2001. Publius et la nature humaine. *Revue Française d'Etudes Américaines.* N°87.

—. 1999. Idéologie et tradition chez Mahan. In *L'évolution de la pensée navale*, ed. Hervé Couteau-Bégarie. Vol.7. Paris : Economica.

Samuels, Ernest ed. 1973. *The Education of Henry Adams.* Boston: Houghton Mifflin.

Scanlan, James. 1951. The Federalist and Human Nature. *The Review of Politics.* N°21.

Sumner, William Graham. 1934. *Essays of William Graham Sumner.* New Haven: Yale University Press.

Taylor, Alan. 2005. Jefferson's Pacific: The Science of Distant Empire, 1768-1811. In *Across the Continent: Jefferson, Lewis and Clark, and the Making of America.* Eds Seefedt, Douglas, Jeffrey L. Hantman & Peter Onuf. Charlottesville: University of Virginia Press.

Tucker, Robert W. & David C. Hendrickson. 1990. *Empire of Liberty. The Statecraft of Thomas Jefferson.* New York: Oxford University Press.

Vincent, Bernard. 1999. *La destinée manifeste. Aspects idéologiques et politiques de l'expansionnisme américain au dix-neuvième siècle.* Paris: Ed. Messene.

Weinberg, Albert. 1963. *Manifest Destiny.* Chicago: Quadrangle Books.

Whitaker, Arthur P. 1954. *The Western Hemisphere Idea.* New York: Ithaca.

Wright, Benjamin. 1949. The Federalist and the Nature of Political Man. *Ethics.* N°2.

PART IV:

POPULARISING SCIENCE

THE FIRST DINOSAURS AND CHANGING MUSEUM PARADIGMS IN AMERICA

MARK MEIGS

The first dinosaur in the Academy of Natural Sciences in Philadelphia, *Hadrosaurus foulkii*, arrived in 1858, preceding the arrival of Charles Darwin's *On the Origin of Species* by only a year and preceding the arrival on the American natural history scene of the great late nineteenth century rivals of paleontology, Edward Drinker Cope (1840-1897) and Charles Othniel Marsh (1831-1899), by less than a decade. The combination of that book and that pair of avid scientist/collectors and the sheer size of the dinosaurs they collected would carry American paleontology first into a bitter contest between museums that could display those beasts and then onto a world stage where adventure novels and fantasy movies have dramatized the struggles of pre-historic beasts and intermingled them, inevitably if unscientifically, with the struggles of men.[63]

The images of pre-historic men facing pre-historic animals in ways that defy the geological and evolutionary record start with Sir Arthur Conan Doyle's novel of 1912, *The Lost World*. Edgar Rice Burroughs published *The Land that Time Forgot*, in 1918. Merian C. Cooper, using these models, but also apparently inspired by a dream of a giant gorilla on the rampage in New York City, created a serialized version of *King Kong*, published in 1932 as publicity for the movie he directed of that name, released in 1933. For the writing, Cooper collaborated with Edgar Wallace, the hugely successful English novelist of the time, who died during 1932. Also deserving credit for the King Kong book is Delos W. Lovelace, a journalist and writer of many popular books from this time till the 1950s. Who actually was responsible for what image in the Kong story is difficult to say, but it is likely that the drama of primitive men facing gigantic, if extinct, species while under the eye of modern men, who would of course eventually prevail, appealed to all these masters of popular entertainment. The image appeals to this day even in venues more scientific than popular movies. A recent issue of *National Geographic* (December 2008, 32), contains a small article on "Fossils" by Helen

Fields, that dramatizes the size of a newly discovered pliosaur fossil by showing its fleshed out profile in a menacing position near a frogman's profile.

The confrontation of men with nature has become a well-studied subtext of natural history museums. In her deliberately provocative book, *Primate Visions*, of 1989, Donna Haraway described the science made visible in the dioramas, so popular in natural history museums in the beginning of the twentieth century, as more than descriptions of various animals beautifully preserved and presented in their scrupulously reproduced natural settings. The specimens, she pointed out, were often arranged in family groups with the male gazing out of the glass case of the diorama confronting the viewer. At times, there was only a male, always a large specimen, always posed dramatically. Her principle example was the handsome silverback gorilla at the American Museum of Natural History in New York killed by the naturalist/big game hunter/conservationist/ taxidermist /designer of dioramas, Carl Ackley (1864-1926) who worked for the Field Museum in Chicago and then, after 1909, for the American Museum of Natural History. This male gaze, she maintained, was a challenge between whatever member of the animal kingdom was presented, and the viewer. Would the human or the animal prevail in this contest? She went on to say that some dioramas preserved for all time the moment when the animal last, and perhaps first, looked at an armed man, the man who was about to shoot it. She described the documentation around the silverback gorilla to demonstrate this point. The same observation can be made with the documentation around the dioramas of African and North American game shot in the same period for the Academy of Natural Sciences in Philadelphia. Along with documenting the environment in which the animals were found, part of the science of expeditions to bring back specimens, was to document the exact circumstances of the shooting specifically for the reproduction in dioramas. Big game hunters, who always wanted to hunt the best trophy, often the largest and most dangerous male available, could serve science while memorializing their adventure.[64]

What Haraway described happened at the beginning of the twentieth century. She analyzed this aspect of science museums by saying that Darwinian theory, popularized by Herbert Spencer as survival of the fittest and then applied to a host of social and historical situations, notably imperial imperatives by John Fiske (1842-1901) and Josiah Royce (1855-1918), had inevitably found its aggressive masculine way into museums. The subtitle of her book, "gender, race and nature in the world of modern science," announced her thesis quite clearly. There were a good forty years, however, between the publication of *The Origin of Species*, and the creation of displays of male, white supremacy that Haraway saw in dioramas. During that forty-year period, it was dinosaur displays, displays in other words of creatures that could not possibly ever have met a human, that led the way towards this museum drama of confrontation. The size of the beasts, and, once assembled, their dominance over whatever space they occupied, along with the early consensus that many of them walked on their hind feet, forced this confrontation between men and dinosaur even among people who resisted Darwin's thesis and whose science was otherwise cautious. A dinosaur could not occupy a case of specimens for classification like other fossil remains awaiting comparison and classification. Once it was assembled it stood alone challenging all comers, and especially men.

In 1858, however, when the first dinosaur arrived in Philadelphia, this skeleton washed in on the long high tide of Joseph Leidy's distinguished career that was formed in another scientific age with other preoccupations than either Darwin's theory of evolution or Herbert Spencer's more confrontational expression of it as "survival of the fittest." The biological paradigm of which Joseph Leidy (1823-1891) was a master was that of identifying, naming and categorizing. It implied certain practices and standards, and a certain deliberate way of illustrating and publishing books and articles that met those standards and distributed the news of discovery and naming to a specialized audience.

Museum displays shared scientific work and discoveries with specialized researchers, amateurs and the curious—categories that were not separated very clearly. Leidy relied, for his collecting, on the good will and cooperation of many amateurs who understood that they needed his expertise to explain their discovery. Popular culture and popular imagination crossed paths with the paleontology done by scientists when a seeming army of alert farmers, railroad engineers, soldiers and whoever else found himself—it was a masculine world—digging on the edges of American civilization and found some strange remains. That "edge" of civilization had little to do with any "frontier" in the sense of an imaginary line moving west with a migrating population. Rural New Jersey, as well as the Dakotas and Wyoming produced fossils. Word of them and the bones themselves would inevitably arrive on Joseph Leidy's desk. He held the title of Professor of Anatomy at the University of Pennsylvania and a membership in the Academy of Natural Sciences of Philadelphia, along with memberships in any number of scientific societies around the world, but it was his reputation as America's foremost vertebrate paleontologist that brought specimens his way.

At first, the new dinosaur remains fit into that old paradigm. William Parker Foulke, the man who uncovered Hadrosaurus foulkii, was an associate of the Academy of Natural Sciences in Philadelphia on vacation in New Jersey where he heard from a farmer about some bones uncovered years before. When he located the fossil site and uncovered more bones he sent them at once to Leidy who included his name as the species of the new discovery.[65] When such discoveries were coupled with the dynamics of evolution, rebuilt into the towering skeletons that still dominate museum halls and then taken up by the fiercely competitive Cope and Marsh, they escaped from naming and categorization. They became both the symbol and example of violent struggles in nature, both dramatizing the fierce qualities of evolution's winners while at the same time becoming a possible example of inadequacies that evolutionary processes could dispatch to the oblivion of extinction.

Leidy's work shows both the methods of his science of naming and his caution. In Figure 1, a drawing of Uintatherium robustum, a creature found in Eocene deposits near Fort Bridges Wyoming, Leidy produced one of many examples of his superb draftsmanship. An associate of Leidy's found the bones in this particular location in Wyoming, in July, 1872. Leidy described the creature and named it in a letter that was read at the Academy before the end of the month and published August 1, 1872. He published lithographs of his drawings a year later in *Extinct Vertebrate Fauna of the Western Territories*. In the meantime, the great dinosaur hunters and rivals Marsh and Cope had also found remains of similar creatures and named them. In Figure 2, in the same publication, Leidy placed some of the bones of his specimens of Uintaherium robustum on the outline of the Dinoceras mirabilis made by Marsh. In this way Leidy claimed that the two animals were in fact the same.

With extreme simplicity and efficiency, Leidy went about his task of naming. He left the impression that had Marsh not drawn the skull, Leidy would have contented himself with reproducing exactly what he had at hand, and not speculated beyond what was required for naming. He suspected that a second of Marsh's finds of that summer, Tinoceras, was also a Uintatherium, but he left that judgment to Marsh who had better specimens to work with.[66] He certainly did not indulge in speculating about the habits of the creature, or try to show it in action, though from the teeth he identified it as an herbivore.

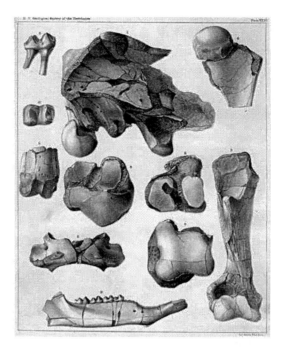

*Figure 1. Plate XXVI from Extinct Vertebrate Fauna of the Western Territories
(1873)
www.ansp.org/museum/leidy/paleo/litho_1873_XXVI.php
Courtesy of the Academy of Natural Sciences, Philadelphia*

From the late 1840s through the early 1870s, Leidy enjoyed a near monopoly on all fossil remains found in North America. His friends and colleagues sent him everything relying on his ability to draw on a great familiarity with published fossils in order to identify new discoveries. His remarkable ability as a draftsman led to the production, year in and out, of beautifully illustrated articles and books announcing the news of paleontology.[67] He, and his more or less educated and informal assistants in the field, worked together on filling in the blank pages of the great book of nature and God's creation but there was a hierarchy in this chain of discovery. Men in the field, whose major occupations were often not scientific, sent their discoveries to Leidy. He published the information in one or another of the scientific journals of the times, acknowledged their assistance. As for God, he oversaw the whole affair, and without interfering too closely with the details, could yet accept credit for it all too, if Leidy were asked.

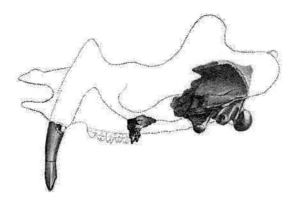

Figure 2. From Plate XXVIII from "Extinct Vertebrate Fauna of the Western
Territories" (1873).
The large fang that Leidy placed in the upper jaw had been found the day
before Leidy's Uintatherium robustus was discovered, and only during a year of
study and comparison of Cope's and Marsh's discoveries, did he place it with this
animal. www.ansp.org/museum/leidy/paleo/litho_1873_XXVIII.php
Courtesy of the Academy of Natural Sciences, Philadelphia

Leidy acknowledged no more conflict between a belief in a Supreme
Being and science—even in the form of Darwin's version of evolution—
than he acknowledged or entered into, conflict between scientists working
in the same field. He reacted violently in 1853 to the accusation that his
writing and teaching revealed an atheism that unfitted him for a chair at
the University of Pennsylvania.[68] Throughout most of his life his lectures
and articles were based on an immediate materialism that let his students
assume his atheism even while he could assure members of the religious
community that his position was shared with John Fiske of Harvard, who
believed that "though Science must destroy mythology, it can never
destroy religion and to the astronomer of the future, as well as the Psalmist
of old, the heavens will declare the Glory of God."[69] Though he was
among the first of the scientific community to applaud Darwin's book and
theory, and caused Charles Darwin to be elected to the Academy of
Natural Sciences at once, in 1859, his thinking remained in a sense pre-
Darwinian. He refused to address the controversy caused by the book and
the paradigm of conflict of struggle Darwin's followers introduced into all
realms of science and representations of science, including those in
museums.

The rivals Edward Drinker Cope and Othniel Marsh had no such
scruples or sense of caution. Cope was determined to prove Darwin at

least partly wrong. Marsh was equally determined to find fossil evidence to prove Darwin right. Marsh's huge collections provided evidence especially for the evolution of horses and birds. Cope published his seminal work whose title announced its confrontation with Darwin, "The Origins of Genera" in 1867 at the age of 27. He never reconsidered his analysis. He admitted that species may be brought about by a natural variety in species that Darwin had noted and given direction by Darwinian "natural selection," but that species must derive from genera, the Linnean category broader than species, but narrower than orders. The origins of genera, he wrote, obey "a system of retardation or acceleration in the development of individuals; the former on preëstablished, the latter on preconceived lines of direction." In other words, God had determined the path by which genera would move forward or backward. Cope was a neo-Lamarckian as well, and so the will and experience of any individual in a species counted in these cycles of acceleration and retardation.[70] The origins of human kind, not unexpectedly, kept Cope from a more thorough acceptance of evolution. "What necessity resulted in man as the crown of the Mammalian series, instead of some other organic type? Our only answer and law for these questions must be, the will of the Creator," he wrote early on in the article. He concluded his article saying

> [...] the genus Homo has been developed by the modification of some preëxisting genus. All his traits which are merely functional have, as a consequence, been produced during the process. Those traits which are not functional, but spiritual, are of course amenable to a different class of laws, which belong to the province of religion.[71]

The positions of these two men lead them into a peculiar and ironic version of dinosaur habits. Marsh, who proposed that the dinosaurs' upright stance resembled the posture of birds and who provided evidence that birds were the descendants of dinosaurs, the creatures themselves, doomed to extinction were slow, propped up by heavy tails that they dragged around behind them. Cope's dinosaurs could be lively, even playful, much closer in fact to the war-blooded, quick *Jurassic Park* creatures of movie and novel that we are familiar with today. Cope's dinosaurs were free from flaws that would prove disastrous in competition because he shared the notion, in the excellent company of Louis Agassiz, that extraordinary catastrophes had from time to time wiped God's creative slate clean in the manner of the biblical flood. In his description of Laelaps Cope, which appeared in the same publication as his "Origins of Genera" (1867), Cope depicted a very agile dinosaur indeed:

Laelaps took enormous leaps and struck its prey with its hind limbs. I say, in describing it, the small size of the fore limbs must have rendered them far less efficient as weapons than the hind feet, in an attack on such a creature as Hadrossaurus; hence perhaps the latter were preferred in inflicting fatal wounds.[72]

At the end of his life, in 1897, Cope oversaw the creation of a series of illustrations of dinosaurs by the artist Charles R. Knight (1874-1953) who rendered this version of Laelaps in full colour (Figure 3). The illustrator Knight produced images that decorate the Field Museum in Chicago, the American Museum of Natural History in New York, the Smithsonian in Washington, D.C., and Yale's Peabody Museum, among others. But this early work, under the influence of Cope at the age of twenty-three shows dinosaurs at their most lively and correct, according to late twentieth century theories. Later under the influence of more convinced Darwinists, Knight put far more doom into far more ponderous creatures. His most famous image, "Tyrannosaurs and Triceratops," painted as part of a large commission for the Field Museum between 1925 and 1929 (Figure 4) shows a compelling, if fictitious confrontation between the horned herbivore and the long-toothed carnivore. It was painted to portray all the hallmarks of late nineteenth century Darwinism, where these two would kill each other off making the way for progress. Since then, Darwin's theory has lost no ground, but Cope's once anti-Darwinian notion of catastrophes overtaking agile dinosaurs has replaced this murky scene of flawed monsters intent on mutual destruction.

What is important here is that the collectors and the objects they collected moved to center stage and their science became invisible in the museums they supplied. Where Joseph Leidy had shown what he did— identifying bones and naming the animal from which they came—Cope and Marsh became adventurers in the field and in publications and used their collections of still objects to demonstrate theories of dynamic change that it is very difficult for objects to show. In fact, dinosaur skeletons became huge and theatrical demonstrations of raw power in support of whatever theory the owner wished to present. The two men bid against each other in the field and argued in scientific journals. Eventually their disputes reached the popular press. The money that allowed them to drive the price of fossils up, and their arguments drove Joseph Leidy and his careful paradigm of naming species off the stage.[73]

Figure 3. "Leaping Laelaps," Charles R. Knight, 1897, American Museum of Natural History (present image taken from "The World of Charles R. Knight" charlesrknight.com/Gallery.htm)

At the same time, the search for fossil remains and the information around it had become, if not more professional,[74] at least more formalized, with the lines of communication and the demarcations of controversy made more rigid. Marsh and Cope, often depicted as rivals, both in the field where they were literally struggling over the same fossil remains, and in scientific journals where they sniped at each other and argued over matters of interpretation and precedence, in fact helped bring this transformation about together. The showmanship of their rivalry over the raw material of science introduced a competitive showmanship into their publications and into the displays of their findings in their respective institutions. Instead of a popular interest in science motivating an interested and diverse population to send the raw material of paleontology up the hierarchy to the handful of experts in a position to publish and theorize about it, the scientists found themselves feeding theory and dramatic images to the popular imagination and paying for the raw material of discovery in a free market of field paleontologists.

This rivalry and the images associated with it came to the attention of Andrew Carnegie in 1897, the year that Cope died. Carnegie is supposed to have seen a cartoon of a dinosaur looking into the window of a tall New

York building. "Get it," we are told he said, while putting a check for $10,000 into an envelope, "Get something as big as a barn." He purchased the remains of Diplodocus in 1898, had them cast in plaster and sent the skeletons around the world with a Latinized version of his name appended to it, carnegiei; he knew very well that he fed the world a popular image of the domination of the United States in science, but also in power and wealth. In fifty years paleontology had become a metaphor for a popular notion of the struggle for supremacy in fields far removed from the modest taxonomy that had been its goal earlier on. In Carnegie's hands, at the end of the nineteenth century, dinosaur paleontology supported a paradigm that explained all human endeavor in terms of struggle and domination even if those in the dominant position, like Carnegie, could also be good stewards of their knowledge and power and spread the benefits of science abroad.[75]

By then, dinosaurs had escaped from the ambitious Cope and Marsh just as they escaped from Leidy. With their metaphoric usefulness recognized by the great captains of late nineteenth century industry, called often enough titans to give them a mythological glow, but never dinosaurs, dinosaur science moved to the front of the hierarchy of museum displays and helped establish a hierarchy of museums and the scientific rank of the cities that contained them. By now, of course, certain dinosaurs have moved beyond the confines of museums to figure in novels and in movies. The fossil skeletons, once the objects of fierce competitive struggles have become quaint, fleshless relics that refer to the terrifying, magnified beasts animated on film when, of course, in scientific terms, the films rely on the bones. Jurassic Park, the best-known dinosaur book and film yet, tells the story of one more dinosaur escape, this time from control in a late twentieth-century park with a dinosaur theme isolated on an island off Costa Rica. In Michael Crichton's novel the dinosaurs escape to the Costa Rican mainland. In movie and novels, freed from scientific restraint or probabilities, the extinct creatures can wreck havoc with our imaginations and confront humans and human society in any number of sequels.[76]

In the Cretaceous and Jurassic periods in which they lived, of course, such contact with humans was impossible, and it was equally impossible in the science of Joseph Leidy. Scientifically such a confrontation is no more likely today, when fiction uses bio-technology to create a world, than it was in the nineteenth or early twentieth centuries when fiction used the notion of un-mapped parts of the globe to lend credibility to the discovery of a cretaceous island far from trade routes or a lost valley in the cartographically blank parts of Africa. Among the many articles about the Jurassic Park movies are a number of serious objections to the science presented in Michael Crichton's fiction that resurrected dinosaurs in modern times. When Sir Arthur Conan Doyle wrote his dinosaur book, in 1912, the first in this long line, there was no longer that much blank in Africa or anywhere else. Implausible as it was, the image of that confrontation between the largest creatures to have walked the earth with our own species has been desired by scientists, writers, and museum directors all. Without Darwin and the changes in museums that happened after his book, it would be hard to imagine the compulsion for witnessing this contest that could never take place. *The Origin of Species* brought in a whole constellation of changes to museums that give a kind of sense to that confrontation. But the size of the creatures themselves first moved them out of the old display cases arranged in taxonomic order putting them in the center of halls where museum visitors confront them head on to this day. (Figure 5).

Figure 5. Diplodocus carnegiei, as he, or she, confronts visitors at the entrance to the Galerie de Paléontologie at the French National Museum of Natural History, Jardin des Plantes, Paris.

POPULAR EVOLUTIONISM
AND THE ETHICS OF PROGRESS

RICHARD SOMERSET

Originally, evolution meant something like "unfolding."[77] Taken in this sense, the "evolution of life" means nothing more than the stages by which the economy of natural species reached its current equilibrium. It was not Darwin, nor even the historicist nineteenth century in general, that created an interest in this subject: the story of how creation took shape has been a commonplace theme of mythological narrative in all times and places. Indeed, the six days of creative work described in *Genesis* might be understood as a form of distinctly pre-scientific evolutionary story-telling. So the characteristic innovation of the nineteenth century was not exactly the creation of a historical discourse of evolutionary becoming so much as its transposition to an empirical register.

In the mythological or cosmological version of the story of creation, there had been no reason to limit the account's argumentative or narratological scope exclusively to the evidence of the senses, or for that matter, to the evidence of revelation. Any argumentative content and any narrative devices that enhanced its appeal were willingly and readily used. It was this intellectual approach that started to change in the eighteenth century, as natural philosophers sought ways of doing the work of traditional cosmology according to the emerging empirical standards of the natural sciences. Being scientific now meant limiting oneself to the empirical facts and eliminating speculative content. But how was this to be done for "the history of life"? What sort of empirical facts were available to this discipline, whose object of study was the "lost world" of a past that had had no human witnesses? All that was available was the fossilised remains of prehistoric animals and plants, objects whose very status was highly controversial.[78] Moreover, the remains were necessarily fragmentary, and naturalists could never hope to have anything like a complete fossil "summary" of the entire story of life on earth. So even within a strict empirical tradition, there would always have to be a place for speculation. However closely the scientific discipline managed to approach that abstract ideal of an account based upon a complete record, a variety of inherent empirical difficulties condemned it to retain something of the character of the classical cosmology: a practice that continued to produce historical narrative rather than natural laws.

As palaeontology emerged from an earlier intellectual model built around pre-historicist natural history and pre-scientific cosmology, there was much to reinvent. The immediate practical problem lay in the difficulty of empirical access, so to speak, to the past; but there was also an important systematic difficulty in the question of how to tell an empirical story of the past. How was the discipline to manage the interaction between empirical credentials on the one hand and narrative coherence on the other? A good story is compelling, but it might also undercut the central effort of an emergent empirical discipline to distance itself from what it wanted to present as the abusive practices of an earlier age. This tension at the heart of the emergent science of the natural past made the construction of a disciplinary identity a complex matter. Much would have to be carried by the capacity of its leading authors to produce narratives capable of harnessing ambient ideologies and to put them to work as part of their argumentative strategy.

The key ideological component for this domain is the notion of progress. This had not in fact always been a sure value. Classical and medieval cosmologies had typically seen the history of the world as regressive, receding from early glories to increasing corruption and baseness, albeit with the expectation of ultimate redemption in the Christian model. But as cosmology was empiricised in the eighteenth century, and so moved towards the historical science we know now, practitioners were *forced* to look for naturalistic patterns in the process. Empiricisation would inevitably tend to divest the natural space of the enchanted character which mythological cosmologies could and did attribute. But the ideology of progress could supply a replacement by investing the course of natural history with a deeply inscrutable sense of purpose. The mysteries of progress thus replace the mysteries of creation.

It was the supposed inherent progressiveness of nature that would ultimately convince the Victorians of the acceptability of Darwinism.[79] The idea appealed because it fitted with a certain liberal outlook that was increasingly used to justify British imperialism as a force for progress around the world. Britain was promoting by policy what could now be presented as the underlying tendency of Nature herself. But on the other hand, the same attitude caused potential embarrassment since, if the inferior classes and races turned out to be perfectible—as the theory suggested—then it would only be a matter of time before the very fabric of society would be torn apart by improved underclasses no longer willing to accept their lowly hereditary station. And if there is one thing as dear to the liberal spirit as the desirability of gradual progress, it is of course the sacrosanct nature of the established order.

Victorian liberals therefore had to work out what to do with organicist progressivism. Progress was useful and attractive, but it had to be suitably controlled. It provided a useful echo for the self-improvement ethos, but it was also important that self-improvement *not* be taken to imply a dissolving of the class barriers that defined the social order. This is what we might call "the liberal conundrum." How are we to depict society as inherently and organically progressive, without thereby undermining the stability of its defining structures?

As life science took on its modern form, so the History of Life genre was able to emerge as an attractive and popular window on a domain that married scientific empiricism and historical narrative. In such texts, the ways of dealing with the narrative of the History of Life are also ways of addressing the ideological issues associated with progress and progressivism. They mix empirical evidence and historical narrative to arrive at a general account of progress: its dynamics, its operation, its impact, its implications. The three models we are going to compare—produced by Buffon in 1779, Robert Chambers in 1844, and Arabella Buckley in 1882—are all different in their conceptions of the progressive course of nature over time as manifested in the successive forms of living things, but they all propose a separation between, on the one hand, a sort of *voie royale* occupied by a select elite responsible for the dynamic thrust of change, and the general mass of creation, on the other hand, which exists merely to fill out the available geographical and taxonomical space and thus to complete the universal order of things. In our study of the ideological content of the History of Life genre, we will therefore focus on the form and the tendency of the dynamic thrust suggested by each author's narrative, and in particular on the mechanisms of inclusion and exclusion. This should enable us to perceive an example of how appeals to science could help the Victorian establishment to maintain a precarious balance between the apparently contradictory values of progressive liberalism and those of hierarchical imperialism.

Buffon's *Epoques de la Nature* (1779)

The first of our examples is not Victorian, nor even British. Georges Leclerc, comte de Buffon, was perhaps the first naturalist to offer a continuous narrative in a purely naturalistic mode that attempted to summarise the story of all the successive ages of life on earth. The account was not at all dependent on the notion of species transmutation.[80] Instead, its narrative dynamic is derived from what we might call a "cooling earth scenario." Buffon speculated that the earth had been formed as an incandescent ball, perhaps broken off from the sun by a meteor collision. Once the globe had cooled sufficiently, forms of life adapted to hot conditions could appear; in subsequent ages, as the earth continued to cool and the climate of its surface to change accordingly, new forms of life could come into being. Each successive "type" was created in the polar region—it was in the coolest part of the globe that the threshold temperatures would first be reached—and then move towards the warmer climes of the southerly landmasses, where they would diversify in response to local conditions into a range of sub-varieties, though always within the limits of variability allowed by the general design of the type.

Buffon's account is thus characterised by two interacting narratives, the one relating to a macro-history of typological progression, and the other to the micro-history of specific variation. This dual narrative scheme greatly enhanced the credibility of the model. By separating out "marginal" material, Buffon was able to reduce nature's chaotic profusion to an underlying core narrative of linear progression. It was this narratological innovation that made the History of Life genre possible, and gave it its unparalleled ideological potential, as we will now see by looking at what became of Buffon's scheme in the hands of popularising authors in the historicist and imperialist nineteenth century.

Vestiges of the Natural History of Creation (1844)

Robert Chambers' anonymously published *Vestiges* was the first effort in Britain to produce a globalising account of the history of life based on transmutation. In a sense, it picked up on the Buffon model, but swept away the typological barriers. Here, life was to be a single continuity, climbing up the scale "from Monad to Man," to use an expression that would become a cliché by the end of the century. In an analogy to the growth of individual organisms, life itself was seen as passing from a primitive to an advanced condition as a result of a spontaneous process like that of the maturation of an individual animal from juvenility to adulthood. The claim was facilitated by embryological studies suggesting that the human foetus "recapitulated" the movement through fish, reptilian and lower mammalian forms before attaining a perfectly human form. This discovery—that ontogeny recapitulates phylogeny—was to take a central place in Chambers' argument, as is plain from the table reproduced below, in which he makes an explicit parallel between the developmental dynamics of typology, chronology and embryology (fig. 1).

Fig.1: Table reproduced from *Vestiges*, 226-7

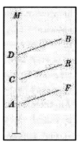

Fig. 2: Diagram reproduced from *Vestiges*, 212.

In the left column are the orders of animals arranged in ascendant chronological order; in the centre the ages of geological strata associated with each of these types; and on the right, the corresponding stage in the development of the human foetus. The implication is clear: the array of types in the animal kingdom are the product of geological ages, and they were produced "organically" by a process of maturation directly analogous to the maturation of the human foetus. The foetus of two months' development has reached the stage of a fish; and after three months it has advanced as far as the stage of a turtle. In fact this parallel between a human foetus of three months and reptiles is not valid, the appropriate comparison being with reptilian *foetuses*; but such quibbles did not bother Chambers who just needed an analogy suggesting that Nature was capable of bridging typological barriers by a process of organic growth. If human foetuses made the leap in reaching maturity, then why should Nature not proceed in the like manner in the production of 'types'?

The argument is neatly summarised in the diagram here reproduced as fig. 2. If an embryo develops only as far as point A, it will come into the world as a Fish at point F; but if for some reason its embryonic development should be prolonged as far as point C, then it will come into the world as a Reptile, at point R; or after further development as a bird at B or a mammal at M.

Chambers' problem was not change itself. His model appealed to the ideology of material progress, ultimately controlled by the wise law-like framework established by God the first mover; a model which, at a pinch, might have been acceptable to the liberal sensibilities of the Victorian establishment. What was threatening was the potential for a Lamarckian reading in which the progressive dynamic is seen as a form of "escalator" carrying life forms inexorably up the scale of being from one stage to the next: not dying out, but moving infallibly upwards.[81] If the organic process underlying the History of Life is interpreted in these terms, then there is an end of all natural hierarchy, and therefore—to the Victorian mind—an end to all natural and social order. Chambers therefore had to distinguish himself from Lamarck not only by explicitly retaining a role for God as grand architect, but also by finding a way to retain a fundamentally stable hierarchy in the natural order of things. Progressive movement was all very well, but its operation must be carefully controlled lest too easy a transgression of categorical boundaries should end up undermining the reality of the natural hierarchy that the barriers were there to define.

Chambers achieved this prodigious balancing act by appealing to an obscure "quinary" classificatory system that had been advanced by a certain Mr. Macleay. Its schematised all taxonomical difference into five related groups whose relationship applied to classes and orders as much as to genera and species. One of the groups was "typical" and the others were all degenerative variants. The details of the scheme are not significant—it was the source of considerable ridicule even for the contemporary audience. Yet the Macleay quinary system was important for Chambers since it allowed him to identify a kind of elite—the "typical"forms— responsible for carrying forward the progressive changes of the organic process; and by the same token to marginalise the remaining four "sub-typical" forms as merely local variants having no bearing on the grand narrative of life.

It is when he comes to treating the question of man's place in the grand scheme of things that the ideological import of Chambers' distinction between "typical" and "sub-typical" characters is most plain. In his initial comments on mankind, Chambers actually leaves the quinary scheme open, suggesting that only one of the five spaces available has yet been filled, and thus leaving space for future variation, and perhaps progress (1994, 276). But when he moves on to consider "the early history of mankind," he reverts to the pattern he had established for other types of animals, suggesting that the relationship between the existing human races can already be mapped in quinary terms. In addition, the hierarchy of the races is determined in the same way as the general typological hierarchy: by differences in degree of embryological development. The differences between human races arise from their different degrees of development, or maturity:

> [All the phenomena of racial difference] appear, in a word, to be explicable on the ground of development. We have already seen that various leading animal forms represent stages in the embryotic progress of the highest—the human being. Our brain goes through the various stages of a fish's, a reptile's, and a mammifer's brain, and finally becomes human. There is more than this, for, after completing the animal transformations, it passes through the characters in which it appears, in the Negro, Malay, American and Mongolian nations, and is finally Caucasian. [...] The leading characters, in short, of the various races of mankind, are simply representations of particular stages in the development of the highest or Caucasian type. The Negro exhibits permanently the imperfect brain, projecting lower jaw, and slender bent limbs, of a Caucasian child, some considerable time after the period of its birth. The aboriginal American represents the same child nearer birth. The Mongolian is an arrested infant newly born. And so forth. (Chambers 1994, 306-7)

It is no accident that Chambers breaks humanity down into five races: Negroes, Malays, Americans, Mongolians, Caucasians; in that order. This application of the quinary classification system conveniently allows him to divide humanity into typical and sub-typical groups—Caucasians on the one hand as against all the rest—and to focus the entire progressive driving force of humanity on the former group. Progress was a spontaneous natural principle, an outlook which satisfied the rational progressivist dimension of Victorian liberalism; but the principle was unequally shared out, and the greater part of humanity could be nothing more than the foot soldiers in the unfolding narrative. The changes of spontaneous, natural progress therefore need not endanger the stability of the established hierarchy. In Chambers' hands, the exclusion tactics pioneered by Buffon had been tuned up to a new pitch of sophistication, flattering the liberal sensibilities of the Mid-Victorian establishment just as the period of maximal imperial dominion dawned.

Buckley's *Winners in Life's Race* (1882)

Within three or four decades of Chambers' attempt to link transmutationist ideas to the liberal paternalist ideals, the place of evolutionary life-history in the public mind had been radically changed by the wide diffusion and general acceptance in scientific circles of Darwin's theory of change by natural selection. The central particularity of Darwin's scheme, which separates it from Chambers' or Lamarck's or the eighteenth century theorisers of naturalistic life-history, was that it was based on ecology and not on cosmology. Where earlier theorisers had sought a mechanical account of cosmological processes capable of producing a naturalistic History of Life, Darwin looked at the local economy of specific interactions, and sought there a principle by which to connect survival to adaptation, and adaptation back to survival. In effect, Darwinism took the transmutatory debate out of the domain of abstract science, and brought it into the domain of historical process: what had once been an essentially cosmological science trying to adapt a discipline based on a-priori reasoning to an empirical age, became a historical science so completely based on the merely historical fact that it sometimes struggled to convince sceptics that it really was a science at all. That, however, is another debate. What is important for us is the fact that as the science moved away from a cosmological to an ecological stance, so the door was opened to popular treatments of natural history that would function primarily as a narrative, and not as a systematic exposé of an abstract theory.

The core concept of these popularising accounts, which enabled them to work both as science and as narrative, was what we would now call "heredity," and at the time would more readily have been termed "descent," as in the example of Darwin's own title, *The Descent of Man*. It provided the fundamental plot dynamic. That dynamic was commonly deployed to moralising ends, typically focusing on a lesson of humility. The scale of the progression of creation was such that man must learn to relativise his own place in it; and indeed, to recognise the part played by beings lower than himself in raising him to the giddy heights he now enjoys. The popular History of Life genre thus made great efforts to "sanitise" the parent discipline in the public eye by focusing on the improving moral lessons it claimed to find in the systems of the naturalists.

It was in the context of this new hereditary outlook that Arabella Buckley produced what is arguably the first History of Life specifically designed to introduce children to the evolutionary outlook. *Winners in Life's Race*, first published in 1882 and subsequently reprinted six times until 1939, never openly presents itself as a defence of Darwinism, nor indeed is it a "beginning-to-end summary" of the ages of life; instead, it is a subtle characterisation of the dynamics of the workings of nature which does no more than hint that the dynamic in question is an evolutionary one. The title sets the tone: Life can be seen as a race, or a narrative of progress; and in life there are winners and losers, the winners being those who have succeeded in carrying the torch on to a higher level of existence than that which they found when they first came onto the scene. Chambers' *Vestiges* used narrative only in support of the theory whose explication remained his central object; in contrast, Buckley put theory on the back burner and let the narrative do all the work on its own. This low-on-theory approach helped to make a captivating story, and allowed her to elaborate on sensitive themes in a way that did not feel threatening.

As we might expect, the policies of exclusion were an important weapon in Buckley's narratological armoury, just as they had been for Buffon and Chambers. But where the narratives they offered were not directly transcribed but merely reconstructed by appeal to arguments and to evidence, Buckley's narrative was much more like an actual story. She was, in a sense, telling the story of species; or, to put it in her terms, the story of winners and losers in life's race. It was in the way of telling that story, and particularly in the way of separating out the winners and the losers, that all the ideological weight was carried.

There are two obvious techniques deployed by Buckley to help her build a narrative that would be suggestive in the appropriate ways. One was a blurring in the core narrative of the difference between "typological series" and "chronological sequence," and the other entailed the concerted effort to find in that same core narrative signs not only of material progress, but also of moral progress. The first was a way of convincing readers of the material possibility of species change, without ever clearly saying as much; while the second was a way of cosying up to transmutation by suggesting that, far from being a blind and amoral material process, evolutionary change was actually Nature's (and ultimately, God's) way of prompting and promoting the growth of moral values.

The use made by Buckley of the typological series, and her tendency to blur it with chronological sequence, is visible right from the first page. The first chapter is devoted to the theme of "the threshold of backboned life," a subject which she treats in thoroughly typological terms by exploring a series of modern creatures that can be considered as filling the first and lowest spaces in the vertebrate category. She places them in order, the movement from the one to the next being a movement from the least accomplished vertebrates towards slightly bigger and better versions of the type. This is strictly a typological series, but we can already sense the emergent underlying chronological narrative. In the second chapter, devoted to the subject of fish, Buckley maintains her focus on modern animals, taking the reader on an imaginary journey from a small stream in which we observe a minnow, to a river where we observe a sturgeon, to the open sea where we observe a shark. These are all modern fish, but they are represented as belonging to different ages: the minnow is the perfected scaly fish, the sturgeon is of an older armour-plated type, and cartilaginous sharks are the most ancient of all. The suggestion is that in our journey down the river, we have also somehow moved back in time. If, having reached the sea, we turn around and look back over the path we have just run as an exercise in typology, we will perceive the same series arranged the other way around, and which now looks like a chronological sequence. Although Buckley does not say so explicitly, this is the course of evolutionary history, a movement that has to struggle against the current to carry fish life from the terrible but lowly shark status all the way up to the diminutive but admirably lively minnow. So with these subtle narrative ploys that meld typology and chronology, Buckley was able to get her readers thinking in what she would have seen as evolutionary terms, without having had to stray into the past, and without having said anything about the transmutation of species.

The same taxonomy-chronology implicit equation is visible in the "picture heading" that Buckley offers up at the head of the opening chapter devoted to the threshold theme (fig. 3). Here, we see a variety of distinctly modern animals arrayed in a distinctly modern landscape, in such a way as to suggest a narrative of continuity. We start with some fish at the bottom; then a frog and a small reptile just above, followed by a sheep in a kind of opening suggested by a curious internal framing device, and then in the centre of this inner circle, a family group of large cattle. Perched on top are a few small birds. What is the nature of this narrative? Is it a design narrative of typological improvement, or can it be read as a chronological pathway out of the past into the present? Technically, the second reading is quite wrong since none of the animals represented here could stand in any other relationship to one another than that of more or less distant cousins; nevertheless, the sequencing invites us to read the narrative both ways. The materialised "threshold" only accentuates the possibility of both readings, even though the sheep cannot obviously be counted a "threshold figure," in either typological or chronological terms, in the development of the vertebrates as a class of beings.

Fig. 3: Chapter heading for chapter 1 in *Winners*

However, the sheep is perhaps a "threshold animal" in another sense that is very important to Buckley, and that is the sense of moral development. The sheep is the first mammal shown, and the defining characteristic of this group is that they nurse their young; a trait that Buckley presents as a sign of the emergence of the moral qualities of affection and attachment. Read in chronological terms, this illustration might be taken as showing something like "the emergence of family values," which is why the central group are shown as a family unit.

In telling her story of emergent morality, Buckley gave pride of place to the birds, whose joyful song and loving disposition towards mates and young made them suitable role-models. On the other hand, certain types of animals turned out to be awkward intruders that had to be written out of the core narrative. The most notable examples of such reprobates are the shark and the ape, the first being an aberrant survivor and the second a freak of nature.

In the context of the guiding Life's race metaphor, the central idea is that the story of Life is a story of chapters, each dominated by a new animal type. Each type has its turn to dominate animal creation and in so doing, it brings some new quality to Life. But when their day is run, these animals must bow out of the race, and pass on the torch to some new and higher life form which, benefitting from the legacy inherited from its forebears, will be enabled to carry the story of Life onwards and upwards towards a new and higher plane. This general picture is established early in *Winners* with the story of the Crustacea, once the "monarchs of the ocean;" but which are represented in the seas of the modern world only by a few shy crabs, reclusive lobsters and their like. Their portion of the race is long since run, so only a few feeble crustacean forms surviving by ruse or by subterfuge continue to exist today. But unfortunately for Buckley, not all ancient types are as meek as the exemplary Crustacea. The most obvious misfit is the shark. In Buckley's scheme of things, which melds type and age, the shark, being a primitive fish type, is an intruder in the modern seas; worse than that, they continue to be big and strong and generally to maintain an anomalous claim to marine domination. Buckley calls them the "*tyrants* of the sea," engagingly hinting at the illegitimacy of this power that has been usurped from the proper typological-chronological nexus.

So how are we to deal with this "aberrant survival," and repair the harm it does to the general pattern of our smooth narrative of typological progress? The answer is to be found in a striking passage in which the continued existence of the shark and the sturgeon—both "old-fashioned fellows" in the world of modern fish—is explained away by the means of a startling comparison to the world of human races and civilisations:

> Clearly the sturgeon is an old-fashioned fellow, as you may see for yourself, when specimens caught in the mouths of our rivers are shown in the fishmongers' shops. I have often wondered, when standing looking at him and at the sharks in the British Museum of Natural History at South Kensington, whether people who stroll by have any idea what a strange history these quaint old fishes have, or how they stand there among the scaly and bony fishes lying in the cases around, just as an Egyptian and a Chinaman might stand in an English crowd, descendants of old and noble races of long ago, whose first ancestors have been lost in the dim darkness of ages, whose day of strength and glory was at a time when modern races

had not begun to be, and whose representatives now live in a world which has almost forgotten them. (*Winners*, 32-3.)

As for the apes, they got even shorter shrift from Buckley. She probably felt that any suggestion of a direct link to man would have been too threatening to her readership, and in fact she went to considerable lengths to block the construction of any such continuity. They are accordingly dealt with early on in the book—*before* a chapter entitled "The Large Milk-Givers which have Conquered the World by Strength and Intelligence"—and her brief comments are designed to minimise the significance of the physiological proximity to man and to maximise the moral disparities.

> [What we know of the apes] teaches us that in their rough way they have developed into strangely man-like though savage creatures, while at the same time they are so brutal and so limited in their intelligence that we cannot but look upon them as degenerate animals, equal neither in beauty, strength, discernment, nor in any of the nobler qualities, to the faithful dog, the courageous lion, or the half-reasoning elephant. (Buckley 1882, 255.)

Once again, the core narrative thrust is shaped by the marginalisation of unwelcome elements. Apes are castigated for being "brutal," "limited in intelligence" and "savage." Like the shark, the ape is aberrant, but in a different way. The shark is out of his time frame, and carries the anatomy to prove it. The ape is an interloper, appropriating an anatomy that is not due to him. The shark is put playfully in his place by comparison to a fallen empire; the ape is put sharply in his place by comparison to a savage. In the place of the interloper, Buckley prefers the dog, the lion or the elephant for their alleged moral qualities, despite the ludicrous physiological hiatus thereby introduced into what was meant to be a smooth, gradualistic narrative.

To complete our survey, it may be interesting to consider the work of a near contemporary of Buckley's, one Henry Knipe, who was writing a little later and for an adult audience, and so felt able to include a substantial amount of material on the evolution of mankind. Knipe in fact produced two different Histories of Life, a popular version presented in the form of an epic poem, *Nebula to Man* (1905), and a more technical version in prose, *Evolution in the Past* (1912); but neither found much of an audience, the first probably being too old-fashioned in form and the second falling between generalist and specialist readerships. But despite Knipe's apparent lack of marketing acuity, *Evolution in the Past* is at least distinguished by a series of exceptionally fine illustrations, executed for the most part by Alice Woodward, and which are useful to us here to round out the story about the treatment of human evolution at the turn of the twentieth century.

In his text, Knipe treated the evolution of Man in the same methodical way as he had done for the rest of the animal creation, if anything granting this single genus a disproportionately ample degree of coverage. He identified the various types of early human species that had been found in fossil form and attempted to fit them into a chronological order. His illustrator provided reconstructions, which clearly show these beings as forming a sequence that bridges the gap between "proper apes" and "proper men." Three types are represented graphically: Pliopithecus, Pithecanthropus, and Neanderthals (figs. 4, 5, 6).

No compromise is made here about the intermediate status of these beings: they look like animals trying hard to be humans—and although none of them is entirely convincing, their efforts do obviously improve! This sequence strongly hints at precisely the view that the Late Victorians had found most threatening: that the moral order of the human world had emerged gradually and directly from a pre-moral animal order, without the intervention of a transcendent Designer. In deference to that sensibility, perhaps, the book's frontispiece tells a surprisingly different story (fig. 7). Unlike the other illustrations, this one is not meant to provide a direct glimpse of some particular stage of the past story of life; it is instead intended to represent, in idealised form, a summary version of the story of life as a whole. So here we find different animal types formally arranged to suggest the appropriate narrative dynamic. The movement follows a clockwise spiral starting from the shadowy birds in the bottom right corner and the serpent emerging from under a stone, moving on to a group of monkeys one of whom holds out a hand in a gesture that implies a possible impediment to the progress of the narrative movement, and a jaguar leering treacherously from behind; then finally, at the forefront, a young man stepping across a mountain threshold, his face turned towards the sunrise and the continuing ascent that we guess lies before him. The figure is more that of a youth than a man; there is nothing whatever of the bestial about him. He is most definitely a Homo Sapiens, and a decidedly Caucasian and hairless specimen at that. Yet he is also meant to be a prehistoric being occupying a "premodern" place in the natural narrative schematically laid out here. This identity is emblematically indicated by the fact that he is naked, and by the fact that he carries an incongruous stone axe or tomahawk in his hand.

Figs. 4 & 5 (top): Stage one (left) and two (right) in the development of Man
according to Knipe (1912)

Fig. 6 (bottom left): Stage three in the development of Man according to Knipe.

Fig. 7 (bottom right): Knipe's frontispiece.

These images speak for themselves. In the detail, Knipe's evolutionary story is a gradualist account of humanity emerging from animality; but as soon as we turn to the attempt to characterise the story as a whole, we immediately revert to a formulation that leaves out the intermediate stages and focuses instead on the leadership of the Caucasian type. As in all whiggish history, the course of the narrative and its dynamic are determined by the end the author has in mind; and that end always has something to do with 'us.' This representation is about power—the power to set the course of history—and there is no space in it for anything other than a distinctly Caucasian type.

This brief survey has covered a wide range of authors and a period of nearly one and a half centuries. In it, we have seen the narrative logic of the History of Life refined as the evolutionary paradigm was fleshed out. As the grand narrative shrank towards a single, unified continuum of progression, so extra pressure developed to address the other side of the equation in that essential tension which characterises the liberal outlook: the need for a stable hierarchy based on difference. The credibility of the emerging narrative as both meaningful and acceptable science rested to a large degree on its capacity to address this tension between progressive movement on the one hand and a stable hierarchy of irreducible ontological difference on the other. All the way down to the early 20th century and beyond, this imperative forced the incorporation of some version of Buffon's split narrative; and each of the authors we have considered had to develop particular techniques for carving out a distinctive space for the non-players, somewhere off the plotline of the central narrative thrust.

In the earlier forms of the developmental model, the division between the mainstream and the marginal stream remained wide enough to exclude the hope of "redemption" for savages in the naturally progressive course of things; but in Darwinian evolution, "redemption" via effort and improvement became theoretically open to all. In order to avoid threatening the liberal establishment's taste for hierarchical order, authors of popular Histories of Life were therefore likely to seek ways to maintain the former assumptions of inequality even at the cost of losing out in narrative coherence. We saw hesitations of this kind in both Buckley and Knipe.

Paternalistic liberalism in the colonies struggled with the same doubts: in theory, the inferior races were amenable to European improvement, and individuals making such efforts were encouraged; but in practice it was unthinkable that any such improvement might finally lead to equality. Stories like Buckley's were created not only to make a credible story about how the taxonomical barriers could be broken in a grand narrative of natural progression; but also as a way of maintaining certain underlying barriers to safeguard a hierarchy of class and race that middle class Victorians were not yet ready to submit to the action of natural processes.

Paradoxically enough, then, in the hands of establishment popularisers, the science of change was a powerful tool for the channelling and the limiting of change. Despite the notorious fears that were raised by the Darwinian scheme of nature, these popularisers were already adept, even before the publication of *The Origin of Species*, and even more so thereafter, at tailoring the message of their emergent science to suit the conflicting aspirations and concerns of a liberal-minded, empire-building establishment.

Works Cited

Bowler, Peter. 1988. *The Non-Darwinian Revolution. Reinterpreting a Historical Myth.* Baltimore and London: John Hopkins University Press.

Buckley, Arabella. 1882. *Winners in Life's Race; or, the Great Backboned Family.* London: Macmillan.

Buffon, Georges Leclerc, Comte de. 1988. *Les Epoques de la nature* [1780]. Paris : Editions du Muséum national d'Histoire naturelle.

Chambers, Robert. 1994. *The Vestiges of the Natural History of Creation, and Other Evolutionary Writings* [1844]. Chicago: University of Chicago Press. (Facsimile edition.)

Gillispie, Charles Coulston. 1959. Lamarck and Darwin in the History of Science. In *Forerunners of Darwin, 1745-1859.* Eds. Glass, Temkin and Straus. Baltimore: John Hopkins Press.

Knipe, Henry R. 1905. *Nebula to Man.* London: J.M. Dent & Co.

—. 1912. *Evolution in the Past.* London: Herbert and Daniel.

Rudwick, Martin. 1976. *The Meaning of Fossils. Episodes in the History of Palaeontology.* New York: Neale Watson Academic Publications.

—. 1992. *Scenes from Deep Time. Early Pictorial Representations of the Prehistoric World.* Chicago and London: University of Chicago Press.

PART V:

SCIENCE AND SPIRITUALITY

THE EDIFYING FOLD:
DISCOMFORT WITH TECHNOLOGY
IN THE JESUITS' LETTERS FROM AMERICA

FRÉDÉRIC DOREL

> I visit the Indians in their wigwams, either as a missionary, if they are disposed to listen to me, or as a physician to see their sick. When I find a little child in great danger, and I perceive the parents have no desire to hear the word of God, I spread out my vials: I recommend my medicines strongly. I first bathe the child with a little camphor; then taking some baptismal water, I baptize it without their suspecting it – and thus I have opened the gate of heaven to a great number (De Smet / Chittenden & Richardson 1905: 1, 185).

We are all familiar with the narratives of the missionaries or of their observers recording their abusive use of science for apostolic purposes. Examples abound, among them this excerpt of a letter by Jesuit priest Peter-John De Smet, dated 1839 from the US Rocky Mountains. However, I would like to go beyond quotations such as the above, which now sound—to me—somewhat trite and limited. I would like to delve more deeply into the nature and the reformulation of the various relationships the Jesuit missionaries maintained with science and technology, a viewpoint which should shed some light on the complex history of both the missions, and the conflict between the Roman Catholic Church and science. I will try to show that parallel to their apostolic strategies the Jesuits have always been deeply influenced by European science. Their accounts of newly discovered cultures and of their own interaction, nurtured with humanism, Baroque, and the French Enlightenment, are also the result of their scientific perception of the environment.

At the same time, it seems fundamental to take into account the influence of the Counter-Reformation and its baroque setting. Seventeenth century Catholicism derived part of its energy from the principle of the proximity of opposites, of inversion, of reversals. The Jesuits stood on frontiers as well as at the meeting points of discipline and freedom, of Roman Catholic centralism and universal scattering, of obedience to the Vatican and political modernity, of the sacred and the profane, of Ancient and New Worlds, of Christians and pagans, of darkness and enlightenment, of error and truth. The Jesuits were—and in a way still are—the apostles of the paradoxical and dramatic metaphors illustrated by French philosopher Gilles Deleuze in his 1988 work *The Fold: Leibniz and the Baroque*. Deleuze describes the fold as a structure of infinity, of division, the unfurling of the inverted lining from the inside. The concepts of bending, flexion and subjection also stem from it. No wonder the Jesuits have rushed headlong into favouring the development and blooming of history and space since the sixteenth century. As discoverers with enterprising and curious minds, they contributed to the deciphering, the coming together and the re-creation of the world, using an assortment of scientific methods displaying their pragmatism, not in their principles, but in their movements. Stubbornly interpreting differences as the promise of a much stronger sought-after identity, they came to master the art of turning confrontation into complementarity, disparagement into asset, failure into glory, the other into oneself. Thus the more different the other was, the more akin he was to them. The uniqueness of Man was then asserted, in a world that was turned upside down compared to the ancient Ptolemaic system.

My point here is to show that their doubts caused the Jesuits to drift away from the colonial applications of the European science, in a strange turnaround. Father De Smet suddenly became suspicious of the collusion between his preaching and the technological impact of Europe on the populations whose unavoidable decline he recorded on a daily basis. Yet it is to be noted that his behaviour should not be linked to the dispute between the Vatican and science on the issue of modernism, but with a deeper and more painful consideration of the tragic impact of Europe on the world.

Science and techniques as apostolic tools

From the sixteenth century up to the nineteenth century, missionaries referred to America as the most disheartening and hopeless of destinations. Here is what an anonymous eighteenth-century Jesuit author wrote in the foreword to the *Edifying and Curious Letters of Some Missioners*:

> The Memoirs from America display for the curious reader objects, which are dramatically different from those of the Levant missions. [...] Constantinople, Syria, [...] the Kingdoms of Persia and Egypt still reveal some evidence of their late splendour [...], all things reminding the reader of the industry, the wealth and the munificence of their first dwellers. On

the contrary, America does not display anything but lakes, forests, bad lands, and Savages.[82]

Thanks to the principle of the proximity of opposites, we will not question the Jesuits' appeal for these supposedly hostile places and peoples. In the face of such painfully odd singularities, the missionaries got organised and adopted a number of strategies built on the phenomenal technical superiority of the Europeans over the Americans at the time. In fact, the Jesuits viewed science as a quick and universal propagation tool, not to be forfeited to competition. As an example, here are four technical and pedagogical illustrations of Christianity:

• The command of written and spoken languages had an extraordinary magical power, in order to negotiate with the chiefs and the shamans, and to gain their protection to address the tribe.

• Music constituted another privileged means of communication with the natives as well as for communication between the natives and Heaven. The impact of musical instruments, which sing as many celestial voices, is spectacular.

• Engravings and paintings served as pedagogical tools recalling the scenes taught. One can notice the taste of Ignatius for edifying imagery as reflected in Jesuit architecture from Europe to Latin America. That visual profusion was to enable the Jesuits, as Maxime Haubert put it, "to teach invisible mysteries to the natives through visible miracles" (1967, 133).

• The European techniques in everyday life such as iron tools, guns, industrial beads, and so on, appeared superior to the natives' traditional tools. Thus the plough summoned plants; and thought transfer was possible through writing. From seventeenth-century Quebec, Father Brébeuf described the arrival of the clock:

> They thought that the clock could hear, mainly when for a joke, one of our Frenchmen shouted at the last stroke: It is enough now, and at once it stopped. When it rings, they say that it speaks [...]. They ask us about what it eats; they spend full hours in front of it, [...] in order to hear it speak (St Jean de Brébeuf 1957, 33).[83]

These apostolic swindles did not seem to represent moral obstacles for the Society of Jesus for whom all means were respectable as long as the conversion of the Natives was achieved through the explanation of the Great Clockmaker's work. "We have to hook them, even with a twist. Every invention is useful to charity, since it becomes divine in our hands" stated Nicola Mastrilli-Duran from Paraguay (Lacouture 1991-92, 2, 245).

Medicine was also an efficient tool. The natives decimated by old-world epidemics, mainly smallpox, turned to the missionaries. Heroes of civilisation, the Jesuits healed, vaccinated, and sometimes cured the sick. In the second half of the nineteenth century, they managed to vaccinate and baptize the natives who dwelled in the missions. Almost all of them survived. The news was spread and the prestige of the Catholic faith was dramatically strengthened overnight. The Jesuits were then regarded as powerful healers, in contrast to the discredited shamans. We should not forget that, even in a context of mutual accommodation, the shamans were usually depicted in the missionaries' writings as dishonest pretenders, disregarding their scientific knowledge, which was subjected to constant battering. In the meantime, the missionaries themselves claimed to dramatically take on the role of the ancestors. That was the confrontation of two magical thoughts, two epistemological systems.

The encyclopaedic journey

What we know about the work of the Jesuits is what they wrote. Every year most of them had to send an activity report to Rome, thus switching from watching to writing. Their accounts were the product of their immersion in the native world and of their ability to draw the attention of a large number of passionate readers. The *Jesuit Relations* in the 1600s, the *Edifying and Curious Letters of Some Missioners* in the 1700s and the *Annals of the Association for the Propagation of the Faith* in the 1800s and 1900s were the printed, translated, and widely circulated mirror images of the numerous and various observations of the priests, in the year following their writing. These texts stood for an unchallenged propaganda mechanism for the fighting Roman Catholic Church challenged by the Reformation and later by various revolutions and republics. They were also major media and financial tools. Two traditionally distinct genres— the classic and the curious—merged in those reports: the travel account and the spontaneous ethnographic description, which depicted supposedly non-scientific cultures and contributed to the European perception of the world. Most readers were of course the faithful, as expected, but there were also potential donors or inquiring minds or scholars attracted by the sudden proximity of such delightful strangeness. Conventional hagiography was also present, as the edifying aspect: apostolic failures, though never completely concealed, were often transformed into triumph of missionary devotion.

In the eyes of the Christian scientists of the time, wildlife surveys were both exotic and useful, since esthetical bedazzlement was an additional means of paying homage to the Creator, which was certainly part of the pedagogical strategies in the Jesuit schools. Every single aspect of Creation was liable to be listed, indexed, catalogued. Not a single nook was to lie fallow in the colonial and encyclopaedic inventories, which focused on the exploitation of apostolic and material resources. Today, that dissection of reality, that organisation of *wilderness* through lexicological accuracy show us how, century after century, Europe has moved from the description of peculiarities for cabinets of curiosities to encyclopaedic typologies, from gathering materials to collecting collections with the development of some sort of knowledge management. Their surveys obviously reflect their scientific approach to the other's cultures, when metaphors and comparisons uproot the object from its wild proliferation. The missionaries' methods were spontaneous and consequently inductive, starting from the observation of facts with the aim of drawing up rules. Gradually over the seventeenth century, Man was no longer only physically described, but also socially analysed: here were the beginnings of anthropology.

Today it is undoubtedly difficult to regard these accounts as indisputable ethnographic information, mainly due to the various intentions which permeate them. But it would most certainly be a mistake to discard them totally, since on the one hand their relevance has now been confirmed by archaeology and ethnohistory, and on the other hand they uncomfortably settled various cultural transfers between Christianity, Native American and European identities, between polygenist theories and references to the Old Testament. Even if the Jesuits hardly ever acknowledged any critical distance with the European values of the time, it is now possible to read between the lines their embarrassment, their changing point of view, and the reformulation of the initial apostolic projects, faced with the mysterious beauty of the world. No cultural relativism though: it was too early, their perception of differences was different from ours, and we should not forget that for the Catholics of the time, Christianity was not only a religion, but also an absolute truth. But those missionaries addressed Europeans, and their accounts gradually influenced Roman Catholicism with their slight cultural decentring after years spent amongst the Hurons or the Guaranis. The missionaries' experience of the wilderness started a revolution in the European way of reading the world.

Folds and withdrawals in Catholic and imperial science

In the course of its history, the Roman Catholic Church rejected then adopted the Aristotelian science after Augustine had proclaimed the harmony of faith and reason. Up to modern times theology, the superior science, regarded natural science as an assistant likely to confirm the word of the Bible. Science was to explain the hows of the Universe, and Christianity the whys, and only the second set of questions was deemed essential. Men of the Church monopolized thinking. As Georges Minois put it: "They were the very masters of knowledge" (1990 1, 19). But with time Europe became a society of science and techniques and the people started expecting answers from the scientists rather than from the priests. The conflict grew bitter in the seventeenth century with Galileo, who pronounced the divorce of the Church and science, calling for the autonomy of the latter.[84] The Church then lost control over the development of science. It had to choose to adapt or to dismiss the theories which threatened its doctrine. It chose the second route—a dead end— turning its opposition to change into a principle. The 1864 *Syllabus* was the apex of that way of thinking. In the meantime, as the validity of scientific conclusions could now be proved thanks to the development of measuring instruments, the Church was compelled to acknowledge the autonomy of Science.

In that history of certitudes, a majority of the members of the Society of Jesus, more faithful to their obedience vow to the Pope than to their habits, seemed to forget their taste for scientific research and the very meaning of the *birettum*: their traditional doctoral emblem. However the Church never was a monolithic bloc. The conflicting voices were not necessarily those of outsiders or of heretics or cranks. Some Jesuits, as Georges Minois wrote about Blaise Pascal—ironically their worst enemy—, experienced the "split of the Christian consciousness [...] dedicated to the quest for Truth [...] in a faith which enforced moral norms onto modern Science" (Minois 1990 1, 1). But they were few and far between and they endured a Catholicism, which was carefully emptied of its creativity, suffused with monarchism, anti-intellectualism, colonialist self-satisfaction, and demagogical pietism.

The Society would still nurture scientists. For them science and virtue could still get along in a common quest for some sort of humanistic concern, from Galileo to Teilhard de Chardin, both dismissed yet protected. Ricci, a Jesuit astronomer in China, had concluded that nothing better than science could make Middle Empire dignitaries listen to Western educated missionaries. The relationship between the Jesuit missionaries and the Vatican was constantly strained and even conflicting. Their *adaptation* strategies of compromise were criticized, indeed condemned, as in the Chinese rites controversy in the seventeenth and the eighteenth centuries. Despite their official submission to papal authority, they continued to protect in secret the scientists among their ranks. For the Jesuits, the enemy might not have been science *per se*, but rather the impact of industrial development on science and on the world. It was thus their experience on the ground, their expertise and their ingenuity that led them into conflict with colonial authorities keen on heavy exploitation and opposed to change.

Eventually the cumulative effects of techniques and evangelization posed the Society of Jesus the problem of the cumulative effects of colonisation and mission in the massive intermixture of Europeanization and standardization of the world. Often the Jesuits tried to distinguish one from the other, as in their successive experiments of Reductions in the north of the continent—in the Rocky Mountains in the nineteenth century—as well as in the south—in Paraguay in the seventeenth and eighteenth centuries. In the cradle of evangelical simplicity of a handful of pre-industrial Christian communities, the missionaries longed for the return of the mythological beginnings of the Golden Age, far away from the fast-expanding colonies—in their eyes the worst possible examples of Christianity. Hence beyond their technical apostolic swindling, and notwithstanding their scientific perception of the world, the missionaries seemed haunted with suspicions of modern Science. More painful even: beyond the traditional *pastoral*, deep within the folds where concepts face each other, several Jesuit missionaries allowed their discontent and even their guilt to show. From seventeenth-century Quebec, Fr. Ragueneau wrote to his European readers: "Since the Savages have entered upon the Christian Faith, [...] they were given over to destitution by [...] Our Lord" (Lacouture 1991-92 2, 293-94). Another Jesuit, Fr. Vimont wrote: "It is a terrible pity to see these poor peoples die in front of our eyes as soon as they enter upon the Christian Faith" (Relations, 2, 1643-1644, 2).

From nineteenth-century France, both colonisation and mission developments drew their inspiration from the Lumières, in a combination of revolutionary and Catholic universalities. That situation was painfully paradoxical for the Jesuits, who were the victims of several revolutions, which they had partly and intellectually prepared. In fact they belonged with both the Enlightenment and the reaction, in a Romanticism inspired by the *Genius of Christianity* (1802) by Chateaubriand. But with their practical skills and their assorted apostolic tricks, should they not be perceived as the devout branch of the great imperialistic wave? Several of these missionaries opposed their Christian universalism to colonial nationalism, their faith in a spiritual finality of nature to Darwinian materialism, for the worst, and to the dangerous social twisting of Darwinism by Spencer and Galton for the best. In the same way they would oppose Rousseau's *Discourse on the Arts and Sciences* and the technicalities of Diderot's *Encyclopaedia*. In substance they reminded their contemporaries that the propagation of the faith with the threat of guns, the trick of steam engines, or the help of phrenology, could not be but mere treachery. Hence they laboured on a careful separation of mission and colonisation, of morals and techniques.

In the Rocky Mountains, Belgian Jesuit Peter-John De Smet crossed rivers and canyons. He was the founder of the short-lived northern Reductions. His letters offered European readers magnificent descriptions. In 1841, being faithful to the tradition, he produced a systematic 19-page inventory of the local wildlife, with charts,[85] blending scientific precision with his personal mystic lyricism. The growing despair of De Smet as a witness to the destruction of the native American cultures directed him more and more specifically towards the miracles of nature. In fact, the Rocky Mountains Reductions created in the early 1840s were being swept aside by successive immigration waves as early as 1847. The US communication network was spreading. The Oregon and California Trails were militarized in the 1850s. In 1862 the Mullan Road was completed between the Missouri River and the Columbia River over more than 600 miles across the ranges of the Rocky Mountains. From that time on, the Reductions turned out to be missions. Being part of the growing network they became mere rendezvous points and settlements for migrants, the last trappers and the last Native Americans. They were no longer institutions created to shield converted populations from European corruption; nor were they grounds for common meaning; they were small towns entangled in the violent turmoil of a civilization change. Out of habit the priests and the natives obstinately maintained privileged links but the former could no longer fulfil the latters' solicitations other than by some remediation alternating between benign incapacity and colonial authoritarianism— depending on the missionaries. De Smet expressed his painful feelings: he knew that the natives could no longer survive unless they surrendered to European technology, to the science of agriculture, to imprisonment in reservations in a world of loneliness and machines, and consequently, beyond assimilation, to inculturation and destitution.

In 1852 De Smet, who had previously advocated for science and techniques in order to set the Native Americans free from destitution and from so-called local superstitions, became aware of their gloomy limits and started to express his scientific grief, an "ethnological bereavement," quoting Claude Reichler (2002 40):

> But then, what will become of the Indians, who have already come from afar to abide in this land? What will become of the aborigines who have possessed it from time immemorial? [...] They readily send their children to school; they make rapid progress in agriculture, and even in several of the most necessary mechanical arts; they carefully raise poultry and domestic animals. We may then hope that the sad remnant of these numerous nations, who once covered America, now reduced to earn their bread in the sweat of their brow (for they can no longer subsist by hunting), will find an asylum, a permanent abode, and will be incorporated with all the rights of citizens of the Union. It is their only remaining chance of well-being; humanity and justice seem to demand it for them. (De Smet 1859, 71)

In 1854, he went on: "The drama of populations has reached its last scene at the east and west bases of the Rocky Mountains. In a few years the curtain will fall over the Indian tribes and veil them forever. They will live only in history" (De Smet 1859, 213). As a minister of a minor and unpopular denomination, he was not in a position to publish his observations with impunity. De Smet then used another voice which was apparently free from any Catholic influence, that of Choctaw Chief Harkins. Thanks to the missionary the voice of the other could suddenly be heard:

> The time is approaching, in which we shall be swallowed up, and that, not withstanding our just claims! [...] Our days of peace and happiness are gone, and forever. No opposition, on our part, can ever arrest the march of the United States towards grandeur and power, nor hinder the entire occupation of the vast American continent. We have no power nor influence over the most minute project of this government. It looks upon us and considers us in the light of little children, as pupils under its tutelage and protection (De Smet 1859, 215-16).

Thus, under the cover of a Vanishing American the missionary addressed two groups of possible readers: on the one hand, the whites who could only be reassured by the submission of a native Chief, and on the other hand the natives who were to be invited by one of them, in a tone of restraint, to assimilate violent change. De Smet was a mere reporter of the events but there seems little doubt that the Chief's speech mirrored exactly the missionary's feelings.

In an 1858 letter, De Smet harshly criticized the principle of Manifest
Destiny:

> The storm, which has just burst forth over their heads, was long preparing
> [...]. We saw the American republic soaring, with the rapidity of the
> eagle's flight, towards the plenitude of her power [...]. She ambitioned
> nothing less than extending her domination from the Atlantic to the Pacific,
> so as to embrace the commerce of the whole world [...]. All bent to her
> sceptre [...]. They are styled savages [...], but we may boldly assert, that in
> all our great cities, and everywhere, thousands of whites are more
> deserving in this title. (De Smet 1859, 347-8)

De Smet and a group of Upper Columbia Native Americans at Fort Vancouver
in 1856.

The year 1869 witnessed another attempt:

> By most persons the capacity of the Indians has been greatly underrated.
> They are generally considered as low in intellect, wild men thirsting after
> blood, hunting for game of plunder, debased in their habits and groveling

in their ideas. Quite the contrary is the case. They show order in their national government, order and dignity in the management of their domestic affairs, zeal in what they believe to be their religious duties, sagacity and shrewdness in their dealings, and often a display of reasoning powers far above the medium of uneducated white men or Europeans. (De Smet / Chittenden & Richardson 1905, 3, 1063)

One of the differences between De Smet and the other travellers was maybe that the latter privileged seeing and knowing whereas De Smet favoured meeting: exchange, mutualisation, and accommodation instead of colonial mission. Thus the Jesuit and the Indian, two antagonistic clichés of order and freedom gave us a glimpse of what the nineteenth century did to the world. In 1890 in Washington D.C., once the short-lived "middle ground" swept away, the Frontier was officially closed. The United States would eventually start to discover what they had become as a country. A threatening future had already arrived. It was the beginning of the great American nostalgia.

Thus, after having abused the Native Americans with the supernatural aspects of techniques, the Jesuits eventually turned away from various applications of science. Even though the techniques had fortunately enabled them to vaccinate and help the natives survive, they had also been the prerogative of an aggressive type of progress. That progress did not only concern the Roman Catholic Church—the missionaries' care often seemed inversely proportional to their geographical distance—but mainly their own conception of the community of men. That conception enabled them to understand before the reformers led by Helen Hunt Jackson, author of *A Century of Dishonor* in 1880, the political dealings of Washington, which led to the parcelling of tribal lands. In the meantime it enabled them to challenge science itself by the uncalled-for questioning of the established knowledge.

Today we know that the human and scientific tragedy, which resulted from the bombings of Hiroshima and Nagasaki, has definitely projected the former mistrust in the technical applications of Science into the moral history of mankind. Now we, children of Big Science, have to listen to the voices of the missionaries calling us from within the deep and intricate folds of the Baroque adventure—to use Deleuze's theory. Prior to the conceited success of the Scientist religion at the end of the nineteenth century, and before the opposition it created, several missionaries—like that humble Rocky Mountains Jesuit, himself a worried ambassador of another vain and ethnocentric religion—were able to launch a creative reflection on the past of the "peoples without a history," and into the future of the other peoples.

Works Cited

Chittenden, Hiram Martin & Richardson, Alfred Talbot. 1905. *Life, Letters and Travels of Father Pierre-Jean De Smet, s.j., 1801-1873.* New York: Harper.

De Smet, Pierre-Jean, s.j. 1875. *Lettres Choisies, 1849-1857.* Bruxelles: Closson et Cie.

—. 1859. *Western Missions and Missionaries: A Series of Letters.* New York: P.J. Kenedy.

Deleuze, Gilles. 1992. *The Fold: Leibniz and the Baroque.* Minneapolis: University of Minnesota Press.

Fülöp-Miller, René. 1933. *Les Jésuites et le Secret de leur Puissance. Histoire de la Compagnie de Jésus.* Paris: Plon.

Haubert, Maxime. 1967. *La Vie Quotidienne au Paraguay sous les Jésuites.* Paris: Hachette.

Hunt Jackson, Helen. 1964. *A Century of Dishonor, a Sketch of the United States Governments Dealings with some of the Indians Tribes.* Minneapolis: Minnesota, Ross & Haines.

Jean, Georges. 1994. *Voyages en Utopie.* Paris: Gallimard.

Lacouture, Jean. 1991-1992. *Jésuites.* Paris: Seuil.

Minois, Georges. 1990. *L'Eglise et la Science. Histoire d'un malentendu.* Paris: Fayard.

Parkman, Francis. 1980. *The Jesuits in North America in the Seventeenth Century.* Williamstown, Mass.: Corner House.

St. Jean de Brebeuf. 1957. *Les Relations de ce qui s'est passé au Pays des Hurons (1635-1648).* Genève: Droz.

Thwaites, Reuben Gold. 1896-1901. *The Jesuit Relations and Allied Documents: Travels and Explorations of the Jesuit Missionaries.* Cleveland, OH.: Burrows Brothers Co., 73 volumes.

White, Richard. 1999. *The Middle Ground: Indians, Empires, and Republics in the Great Lakes Region, 1650-1815.* Cambridge: Cambridge University Press.

Periodicals

Annales de l'Association de la Propagation de la Foi, collection faisant suite à toutes les éditions des Lettres Edifiantes et Curieuses. 1825-1885. Paris: Perisse Frères, Vols. VI-LVII.

Lettres Edifiantes et Curieuses écrites des Missions Etrangères. 1781. Paris: Mérigot le Jeune. Vols. VI to IX, *Mémoires d'Amérique.*

Mengarini, G., s.j. 1889. The Rocky Mountains: The Memoirs of Father Gregory Mengarini. In *Woodstock Letters* 18, Woodstock, MD.

Reichler, Claude. 2002. Littérature et anthropologie. De la représentation à l'interaction dans une *Relation de la Nouvelle-France* au XVIIe siècle. In *L'Homme* 164, oct.-dec.

Relations des Jésuites de la Nouvelle-France. 1858. Québec: A. Côté éditeur. Vol. II, 1643-1644.

Motion picture

Joffé, Roland. 1986. *The Mission.* Starring Jeremy Irons & Robert De Niro.

Archives

Washington State University Libraries. Pullman, WA. Holland Library. Manuscripts, Archives & Special Collections.

Photograph: *Father De Smet and a group of Upper Columbia Native Americans who surrendered to General Harney at Fort Vancouver in 1856.* By courtesy of the Washington State Historical Society, Tacoma, WA. All rights reserved.
http://digitum.washingtonhistory.org/cdm4/item_viewer.php?CISORO OT=/indian&CISOPTR=31&CISOBOX=1&REC=17

QUESTIONING THE EMPIRE OF SCIENCE: JOHN MUIR'S EPISTEMOLOGICAL MODESTY

JEAN-DANIEL COLLOMB

Principally known for the role he played in the promotion of the US national park system in the nineteenth century, John Muir was also a man of science who enthusiastically shared in his contemporaries' interest in the progress of scientific and technical knowledge. He was undoubtedly a peculiar type of naturalist, relying on an acceptance of the mystery of nature and of the limitations of human understanding at a time when a good many Americans embraced science and its sundry implementations almost unquestioningly. Hence, as an individual scientist, Muir was laying the ground for a more modest and prudent ecological ethos, which would differ radically from the overconfident and unlimited trust in the potentialities of science and technology that helped foster America's rise to global prominence.

A man of science

Muir's father was a Campbellite. He belonged to the Disciples of Christ, a fundamentalist offshoot of Presbyterianism. As a religious dogmatist, Daniel Muir frowned upon his son's pursuit of knowledge, whether scientific or literary. He considered the Bible to be the only book which it was legitimate for a dutiful Christian to read. As a result he repeatedly attempted to divert John from his educational endeavours. The young man had a very hard time of it but his desire to learn was so strong that he strained to circumvent his father's stern authority. In his teenage years, Muir managed to secretly borrow books unrelated to religious matters from young neighbours living in more liberal households. He was and remained largely self-taught and eclectic in his interests and was always very keen to preserve his intellectual independence. At the university of Wisconsin, which he joined in his early twenties, Muir took great care in determining his own course of study instead of fitting the mould (Muir 1997(a), 140). His was a life which exemplified man's desire for knowledge and the American taste for self-reliance in every walk of life. On his Wisconsin farm Muir had turned out to be a surprisingly precocious inventor of mechanical devices. With little material assistance he had managed to work wonders. In his autobiography he mentions his inventing a self-setting sawmill, then goes on to draw an impressive list of the devices he created thereafter (Muir 1997(a), 122). Throughout his life he never foreswore his dedication to scientific precision and rational forms of thinking shorn of obscurantist influences. Muir considered it his task to discover and record natural laws pointing to an inherent order at work in nature. In the meantime he persistently refused to indulge in what he regarded as obscurantism, the enlightened spirit's nemesis. That is why he mocked the popular success of phrenology he witnessed in the Wisconsin of the mid-nineteenth century (Muir 1997(a), 131). He discarded the pioneers' interest in such fields as parochial obscurantism.

It should come as no surprise, therefore, that when he came of age, Muir went on to become a seasoned amateur naturalist. More precisely, he can be described as "a naturalist-in-the-field," to quote George Canguilhem's portrayal of Charles Darwin (1994, 101). The method of scientific research adopted by these naturalists hinged on a direct contact with nature. Pointing out the need to analyse natural phenomena at first hand, the so-called "naturalists-in-the-field" endeavoured to explore nature as frequently as possible and deduce natural laws from empirical observations. They valued empirical experience more than the knowledge they could gain from books. In many respects, John Muir epitomized the ambitions of the "naturalists-in-the-field." He tirelessly pointed out that it was absolutely necessary for man to enter into direct contact with nature if he wished to comprehend it. Consequently he refused to be confined to a study or a laboratory. In fact he secretly despised those who were, as is shown by a remark he jotted in his personal notes in the late 1870s: "More wild knowledge, less arithmetic and grammar – compulsory education in the form of woodcraft, mountaincraft, science at first hand" (Muir 1999, 55). In the mind of John Muir, science at first hand is indeed the only kind of science worth pursuing. Thus he went so far as to assert that mountains and forests were the best universities and libraries one could have access to. In the very last lines of *The Story of My Boyhood and Youth*, Muir reminisces on the moment when he had to leave his university. He was then fairly sure that his process of learning would not come to an end, since he was about to discover American forests. He took it for granted that, by so doing, he was bound for a world of knowledge: "I was only leaving one University for another, the Wisconsin University for the University of the Wilderness." (Muir 1997(a), 142) Nothing could be more valuable than "the University of the Wilderness," Muir thought. That is why, when J.D. Runkle, president of MIT from 1868 to 1878, asked him to come and teach at his university in the early 1870s, Muir politely declined the offer as he was afraid to lose touch with nature. Furthermore Muir believed that in libraries or lecture-rooms science would inevitably lose its appeal. He did not doubt that books and lessons would only be poor surrogates to a ramble in the forest or the ascent of a peak.

Muir was also unwilling to specialise in one particular area of study. He simultaneously dabbled in geology, glaciology, ornithology and botany, with a marked preference for the latter. After a couple of decades Muir could claim a distinguished record as a self-professed non-professional naturalist. Although he always refused to teach at a university, Muir rubbed shoulders with some of the most eminent scientists of his time. He was linked to Louis Agassiz, the man who held sway over American academic circles for about twenty years until his all-out rejection of Charles Darwin's theory of evolution caused him to lose much of his influence.[86] Agassiz and Muir exchanged several letters devoted to the subject of glaciology. Following this correspondence, Agassiz went on record as saying: "Here is the first man I have ever found who has any adequate conception of glacial action" (Muir 1924, 293). Muir also met Agassiz's main rival, Harvard botanist Asa Gray. Muir got informally involved in the botanical network Gray set up at a national level.[87] Like J.D. Runkle before him, Asa Gray tried to convince Muir to move east, but all to no avail (Muir 1924, 292). Muir refused for the same reasons he had declined Runkle's offer. He met and befriended other renowned scientists. He went on several excursions with Joseph LeConte, one of the first professors to join the newly-established university of California in 1868 and who was instrumental in developing this institution. Muir was also acquainted with Sir Joseph Hooker, Charles Darwin's right hand man, and was even invited to his house in England (Muir 1924, 282). Although John Muir cannot be said to have been a member of professional circles in his own right, he was at once attracted to, and accepted, by many of their eminent members.

Muir turned out to be a proficient student of natural phenomena mainly because scientific observation was an activity he would never relinquish. To him science was the passion of a lifetime. He never stopped wondering how men ought to use nature and live in it. A scientific use of nature was one of his favourite answers. It consisted in contemplating the harmony of nature in order to comprehend it. Muir was driven by an urge to know the world. What he was really seeking was an insight into what Ralph Waldo Emerson once dubbed "the method of nature," namely the harmonious and balanced order which nature invariably attains when left to her own devices. And yet, when it came to determining what the aims of science should be, Muir reached some unusual conclusions which cast him as a strange and isolated figure in the age of progress.

A dissenting voice

From the very beginning Muir would have nothing to do with the growing specialisation in the natural sciences, a trend which was already well underway during the last two decades of the nineteenth century. Until then, there had been, both in Europe and America, a rather porous border between philosophy, poetry and science. More often than not these intellectual activities seemed to overlap and go hand in hand. According to Laura Dassow Walls, it was only reluctantly that American intellectuals endorsed the separation between poetry and science, and the division of the latter into well-defined areas of study:

> Specialization and disciplinary boundaries ultimately prevailed because they created the conditions for an astonishing productivity. But while natural philosophers and poets together hailed the growth of knowledge, they also viewed its consequent subdivision and fragmentation with dismay. (Walls 1995, 7)

This uneasiness about the newly established dichotomy between science and poetry may also help us grasp why the expression "natural philosopher" remained so long in use to designate those who indulged in the study of nature. Only at the end of the nineteenth century did the term scientist in the modern sense come into common usage. By contrast to the new scientist, a natural philosopher in the old sense did not have to strictly abide by well-defined and constraining methods of analysis. As for John Muir, although he was very much interested in new trends, he always felt close to natural philosophy.

It must be added that in nineteenth-century America the influence of transcendentalist thought also made itself felt. Several characteristics of transcendentalism were out of tune with the new scientific age which was looming ahead. The transcendentalists had no intention of adopting what they saw as a cold and distanced approach to natural phenomena, which would imply a ruling out of instincts and emotions. As they were influenced by transcendentalist thought, a great number of naturalists still wanted to heed the beauty of the natural world. Such was the case for instance of geologists in the 1870s, according to Michael L. Smith: "The geologists described their work in terms that revealed strong links between the scientific and the aesthetic appeal of mountains." (Smith 1987, 75). The alliance of science and aesthetics stemmed from the contention that without one or the other, man's understanding of nature would be incomplete. Henry David Thoreau was one of the most vocal opponents of the purely materialistic outlook in science which gave short shrift to aesthetic or poetical considerations. In *The Maine Woods* he bluntly refused to be dragged into such a scheme. Observing a will-o'-the-wisp he made it plain that there was more to it than a mere physical manifestation:

> I let science slide, and rejoiced in that light as if it had been a fellow-creature. I saw that it was excellent, and was very glad to know that it was so cheap. A scientific *explanation*, as it is called, would have been altogether out of place there. That is for pale delight. Science with its *retorts* would have put me to sleep; it was the opportunity to be ignorant that I improved. It suggested to me that there was something to be seen if one had eyes. It made a believer of me more than before. I believed that the woods were not tenantless, but choke-full of honest spirits as good as myself anyday [*sic*],–not an empty chamber, in which chemistry was left to work alone, but an inhabited house,–and for a few moments I enjoyed fellowship with them. Your so-called wise man goes trying to persuade himself that there is no entity there but himself and his traps, but it is a great deal easier to believe the truth. (Thoreau 1973, 181)

To modern readers this may well sound like a rejection of scientific inquiry but it would be more accurate to say that Thoreau was merely taking issue with the latest development in science, and not with science *per se*.

Secondly, transcendentalism was accompanied by an interest in philosophical idealism. Idealism implied that nature was a reflection of higher principles, as Ralph Waldo Emerson put it in his essay tersely entitled *Nature*: "Nature is the symbol of spirit." (Emerson 2001, 35). This was a statement Emerson reiterated several times and with which Muir heartily agreed. The consequence of transcendentalist influence was an obsession with the purported unity of nature and the discovery of it on the part of naturalists. The world was to be equated with a unified whole characterised by harmony and order. In those circumstances, the naturalist's role consisted in proving this statement right. The notion of an underlying unity of nature was foremost in the minds not only of most American naturalists but also of American poets and philosophers. They set about uncovering nothing short of the general principles which dominated the cosmos as Emerson claimed: "Whenever a true theory appears, it will be its own evidence. Its test is, that it will explain all phenomena." (Emerson 2001, 27-28). That is why for many decades holism was the norm among American intellectuals.

In point of fact, the naturalist to whom the transcendentalists' commitment to a holistic view of nature was mainly owed was Muir's childhood hero, Alexander von Humboldt. Humboldt had been exceedingly popular in *antebellum* America and remained a highly respected figure into the last decades of the nineteenth century. Following his long drawn-out tour of South America in the years 1799 and 1800, the German naturalist went on a trip to the United States where he was hailed as a hero and introduced to Thomas Jefferson. He had a lasting influence upon those Americans who were interested in the natural sciences and Muir was no exception. In *The Story of my Boyhood and Youth*, he writes that as a boy he dearly wanted to become a Humboldt (Muir 1997 (a), 129). Later on, in 1867, after an accident at a factory in Indianapolis which left him blind for several days he decided to embark on his thousand-mile walk to the gulf of Mexico hoping to walk in Humboldt's footsteps in the Amazonian forest.[88]

Humboldt expressed most of his ideas in *Cosmos*, his monumental study of the natural world, published between 1845 and 1862. *Cosmos* can be viewed as an attempt at contemplating and studying the world in its entirety. Aesthetic contemplation and the development of rational knowledge were the two motives behind Humboldt's intellectual endeavours. What was striking about his work was that aesthetics and rationality were not deemed to be mutually exclusive, quite the contrary. Instead they seemed to complement each other. In *Cosmos*, the scientific and the literary merge. Humboldt had indeed no intention of making a choice between science and literature. In order to achieve this aim he proceeded thus: he would first explore and observe nature, acting in the manner of Canguilhem's "naturalists-in-the-field." Then he would endeavour to take an unprejudiced look at the natural world, collecting as much data as possible. But to Humboldt, empiricism could not be the whole story. The last stage of the Humboldtian method of scientific investigation served to connect observed facts and measurements so as to contemplate and comprehend nature from a general plane. Empirical though it may appear at first glance, Humboldtian science is in fact aimed at gaining a holistic view of nature. Humboldt wanted to avoid a fragmented perception of nature at all costs. Ultimately, the key central focus is upon the unity of the cosmos. There was no denying that naturalists had to start from empirical evidence but only so as to provide a holistic explanation in the end. Humboldt claimed that mere empiricism did not amount to much. Muir was immensely impressed by Humboldt and enthusiastically took his teachings on board. In the journal he kept during his thousand-mile walk through America, Muir paid tribute to Humboldtian holism: "There is no fragment in all nature, for every relative fragment of one thing is a full harmonious unit in itself. All together form the one grand palimpsest of the world." (Muir 1998, 164). This could be a fair definition of the Humboldtian ethos.

However, as the nineteenth century drew to a close, a gradual shift away from holism occurred in the natural sciences. More and more naturalist practitioners were being asked to take a detached look at nature's economy and choose a specific area of study. In the meantime scientific discourse was gradually being deprived of any reference to aesthetics. This all made for the more fragmented and materialistic approach which Thoreau dreaded, as it became increasingly difficult for scientists to keep abreast of the latest developments in every field of scientific enquiry. Humboldt's combined emphasis upon science and aesthetics was rejected as lacking in precision and rationality. Early twentieth-century scientists tended to regard Humboldt's *Cosmos* as a token of former methods by then clearly out-of-date.

This development marginalised the intermingling of literary language and scientific discourse which was the very definition of Muir's writing style. His prose usually has a particular rhythm to it, which rests on an alternation of scientific discourse and poetical language. In a letter he sent to his editor, Robert Underwood Johnson, while he was writing his first book, *The Mountains of California* in 1894, Muir provided his own definition of what he was trying to achieve: "Read the opening chapter when you have time. In it I have ventured to drop into the poetry that I like, but have taken good care to place it between bluffs and buttresses of bald, glacial, geological facts" (Muir 1924, 287). Muir went as far as to associate taxonomy, rational analysis and empirical observation on the one hand, and a lyrical and metaphorical style on the other. He was as likely to quote William Shakespeare and Robert Burns as Charles Darwin or Asa Gray. The eclectic character of his books' content bears witness to his commitment both to science and literary imagination, in the vein of Alexander von Humboldt. That is why one may easily find tables of measurements, precise accounts of natural phenomena and pieces of poetical eloquence all on the same page written by John Muir. The Sierra Nevada chapter of *The Mountains of California* is a fair illustration of this dual approach. In it Muir tries to give his reader an idea of the impact of glaciers on the formation of the mountains of the Sierra. At first the reader is presented with a long paragraph written in a concise and straightforward style which unmistakably qualifies as a piece of scientific writing (Muir 1997(b), 324). And yet, the writer is loath to conclude his observations without lauding the harmonious beauty of the landscape at hand. All of a sudden clouds take on rich symbolical meaning: "Contemplating the works of these flowers of the sky, one may easily fancy them endowed with life: messengers sent down to work in the mountain mines on errands of divine love" (Muir 1997(b), 325). In a few lines, the writer has substituted a romantic praise of the spectacle of nature for plain scientific explanations. Because Muir did not conceive of science as an activity which leaves no room for human emotions and aesthetic taste, he assumed that to observe and understand nature could not leave one indifferent. In that respect, Muir emerges as an excellent representative of American science as it was before the end of the nineteenth century.

Muir's stylistic intermingling blurs the line separating science and romanticism. For instance, in *The Mountains of California*, he refers to the Sugar Pines as "the priests of pines." (Muir 1997(b), 413). Such metaphors may sound confusing to a reader accustomed to a clear-cut dichotomy between scientific discourse and literary language. It is necessary, however, to make a distinction between content and style. When Muir equates the palimpsest of nature with the Tablets of Law, there seems to be little room left for science:

> The canyons, too, some of them a mile deep, mazing wildly through the mighty host of mountains however lawless and ungovernable at first sight they appear, are at length recognized as the necessary effects of causes which followed each other in harmonious sequence – Nature's poems carved on tables of stones – the simplest and most emphatic of her glacial compositions. (Muir 1997(b), 357)

At first glance this passage sounds merely poetical but on further examination it is pervaded by the principle of rationality. The underlying message is that through the spectacle of natural profusion one may discover the very laws which govern nature.

Such a writing style and such an approach to nature were fast becoming anachronistic in the final years of the nineteenth century. It was being replaced by a positivism which Muir was reluctant to endorse because he deemed it too fragmented a way to deal with nature. He stuck to his initial dedication to Humboldtian empirical holism, underlining the need both for rational thinking and the faculties of the imagination. This is precisely what Muir wrote in an entry in his diary in 1906:

> The man of science, the naturalist, too often loses sight of the essential oneness of all living beings in seeking to classify them in kingdoms, orders, families, genera, species, etc, taking note of the kind and

arrangement of limbs, teeth, toes, scales, hair, feathers, etc., measured and
set forth in meter, centimetres, millimetres, while the eye of the Poet, the
Seer, never closes on the kinship of all God's creatures, and his heart ever
beats in sympathy with great and small alike as 'earth-born companions
and fellow mortals' equally dependent on Heaven's eternal love. (Muir
1979, 434)

Without an acknowledgement of nature's beauty, Muir believed that
science was not worth practicing. Until his death in 1914, he took up
several scientific subjects and always sought to uncover general principles
applying to nature as a whole.

Not only did Muir dissociate himself from the latest developments in
scientific methods, he also came to question the triumph of technical
progress, a by-product of scientific research. After the Civil War,
American society came under a flood of new mechanical devices which
profoundly altered people's habits (Jones 1995, 331). This paved the way
for a marked change in American life as expressed by the domination of
an all-pervasive principle of efficiency. Muir was in two minds about the
new industrial age. He was not comfortable with the kind of society the
United States turned itself into after 1865. If we refer to the work of
American historian Richard Hofstadter, such misgivings mark him out as
an unusual thinker in the American context. In *Anti-Intellectualism in
American Life*, Richard Hofstadter argues that American life is remarkable
for its consistent disdain for the past and its fascination for technical
innovation. Thus, Hofstadter holds that in the 19th century what
distinguished American intellectual elites from their European
counterparts was an almost uncritical acceptance of the dawning
technological order:

> Everywhere, as machine industry arose, it drew a line of demarcation
> between the utilitarian and the traditional. In the main, America took its
> stand with utility, with improvement and invention, money and comfort. It
> was clearly understood that the advance of the machine was destroying old
> inertias, discomforts, and brutalities, undermining traditions and ideals,
> sentiments and loyalties, esthetic sensitivities. (Hofstadter 1963, 239)

Hofstadter goes on to point out that in Europe, a tradition of protest against
industrialism and its mechanical realizations arose, a resistance to change
embodied by the likes of Goethe, Blake or Carlyle (Hofstadter 1963, 239).

Hofstadter does not deny that similar voices arose in America but he
holds that they lacked the scope of their European counterparts. According
to Hofstadter there were only a few isolated American writers who
criticised the reign of utilitarianism and the age of the machine, the most

vocal of whom was H.D. Thoreau (Hofstadter 1963, 240). In *Walden*, Thoreau warns his readers against the dangers of a mechanised existence: "[…] men have become the tools of their tools" (Thoreau 1992, 25). This comment testifies to Thoreau's fear of being enclosed in a mechanised world precluding man's freedom.

In a similar way Muir stands as an exception in the American context as laid out by Richard Hofstadter. His was a paradoxical position to say the least. In his youth Muir was adept at inventing mechanical devices and then proved to be an excellent technician, taking jobs in several factories at various moments of his life. This, however, should draw our attention for another reason. Muir kept leaving these jobs in order to go out into the wilds, despite his employers' entreaties. The most spectacular instance of this tendency in Muir was the thousand-mile walk through America he undertook after an accident at one of these factories. In the journal of his travels, Muir repeatedly expressed his exhilaration at being freed from factory work and at having broken away from the deadening regularity of a mechanical existence. He was often critical of what he suspected to be man's enslavement in a mechanised world. Captivated by science and attracted to technical innovation though he may have been, Muir never worshipped science and technology as such. He consistently refused to be dragged into an enclosed world driven by mechanical necessity and pure rationality. Instead he idealised nature and valued its beauty more than human achievements. He was apt to claim the superiority of nature over civilisation and he often couched his praise of wild nature in terms reminiscent of poets' evocations of the garden of Eden. French philosopher François Dagognet holds that the adversaries of technical progress have always been prone to brandish wild nature as a domain superior to and more fertile than human civilisation. In such a scheme, man's technical feats inevitably pale into insignificance when compared to the originality and benevolence of nature (Dagognet 2000, 35-36). Muir's work is a case in point. In *Steep Trails* he seeks to belittle the steam engine (Muir 1994, 249). Even though the steam engine bore testimony to man's abilities, according to Muir, it was nothing compared with the sheer beauty of the spectacle of nature. This was an unusual statement to make. At the time, the steam engine was endowed with highly positive connotations in the mind of the general public. It was seen as the very embodiment of the new industrial age marching forward on its glorious course. Muir begged to differ. Although of much interest to him, technical progress should not be the sole preoccupation in the minds of men. In other words it was wrong for men to be infatuated with their own achievements when wild nature had higher truths to teach. By refusing to promote science as a new religion and by taking a prudent and even ambivalent approach to the rampant process of technical innovation at work in America, Muir was calling for an attitude of restraint and modesty on the part of his contemporaries.

Epistemological modesty

It is fair to say that Muir was a writer out of step with the gospel of progress. Manifesting itself in the last decades of the nineteenth century, the gospel of progress rested on a belief verging on the religious that by applying reason, developing science and converting scientific knowledge into technical innovation, men would be able to improve materially, intellectually and morally, and achieve happiness. The gospel of progress required a break with the past in the form of tradition, customs and old beliefs which were all discarded as obstacles on the progressive eschatological trajectory of human destiny. At bottom, many of Muir's contemporaries thought that science and technology were the means by which men would subjugate nature for good and bend it to their desires. Such a frame of mind held sway in most parts of the western world during the second half of the nineteenth century. Progress as a secular religion set the tone for an age dominated by scientific hubris. One may even argue that the period was characterised by a fantasy of control and omnipotence. In this context, science was bound to be used for the sole purpose of mastering nature and making it instrumental to material progress. Typical of this perception of science were the great surveys of the American West which were carried out between 1867 and 1879. Donald Worster provides an interesting definition of these surveys:

> The survey demands scientific expertise; it is a project characteristic of a modern nation-state steeped in the perspective of science. The survey is a more thoroughgoing way of taking possession; of establishing empire. (Worster 2002, 203)

Americans were increasingly willing to understand and control their vast territory in order to use and exploit it to the full. Seen in that light, science was just a means to that end.

Likewise the conservationist policies launched by Gifford Pinchot, head of the Division of Forestry at the turn of the twentieth century, were a response to the desire to control the natural environment and exploit it as thoroughly as possible. It should come as no surprise that science underpinned conservationism. Conservation should by no means be described as an early form of ecocentrism. In truth it was yet another manifestation of utilitarianism in a more scientific and rational form. Its goal was to exploit nature for the benefit of men while making sure that they would still be able to do so in the longer term. It would be wrong to regard conservationism as an attempt at halting man's exploitation of nature. It was rather a plan to make it more rational and enduring. Its leader, Gifford Pinchot, had studied for a couple of years in France and Germany in the late 1880s, and when he sailed back to the United States, he imported the new science of forestry which he defines thus in *Breaking New Ground*, his account of the early years of the conservationist movement: "Forestry is Tree Farming. Forestry is handling trees so that one crop follows another. To grow trees as a crop is Forestry" (Pinchot 1987, 31). *The Fight for Conservation*, one of Pinchot's other books, was highly anthropocentric in tone and content. In it, he even takes up Jeremy Bentham's utilitarian motto and makes it consistent with the conservationist rationale: "Conservation means the greatest good to the greatest number for the longest time" (Pinchot 2004, 22). Pinchot's management of natural resources can thus be viewed as a continuation of the productivistic mentality so prevalent throughout the industrial age. It was merely intended to make exploitation less damaging for ecosystems which came under great duress on account of America's buoyant economic activity and its unquenchable thirst for raw materials. Man had to retain the upper hand. As Muir was struggling to grapple with the anthropocentric mentality of his fellow Americans, Pinchot was giving it a new lease on life. At first allies in the 1890s, Pinchot and Muir fell out over the building of a dam in Hetch Hetchy Valley, California, which the former supported whilst the latter condemned it as an "outrageous scheme" (Muir 1997(c), 816). Muir thought that it was necessary to set limits to utilitarianism and to the imposition of a scientific and technical order on nature. He believed that Hetch Hetchy had an intrinsic value and was worth preserving in and for itself regardless of man's opinions and desires. But this he did not say or write, knowing that such remarks would not go down well with American public opinion. He chose instead to lay the stress on the magnificent features of the valley and the moral message its scenery could convey. In the end, Pinchot prevailed and the construction of the dam proceeded.

Pinchot's brand of science was profoundly different from Muir's, which emphasised the notion of discovery and observation, and seldom had exploitation as a purpose. Muir's science required man's unobtrusive presence in nature. His scientific ventures ought to be regarded as visitations rather than conquests. As a naturalist, he was intent on gazing and marvelling at nature, taking great care not to disturb its balance in any way. It seems, therefore, that Muir's mode of scientific enquiry was characterised by what Alain Suberchicot called "epistemological modesty" (2003, 139). According to Alain Suberchicot, epistemological modesty finds its origins in the works of Emerson and Thoreau. They in turn gave rise to an intellectual tradition which continued well into the twentieth century. Epistemological modesty ushered in an acknowledgement of the inevitable limitations of human understanding. One must be careful not to mistake epistemological modesty for a disparagement of scientific activities. It is supposed to prompt men to study nature and do their utmost to comprehend it. In the meantime naturalists have to bear in mind that ultimately part of the mystery of nature is bound to remain uncovered, as if out of scientific reach. Epistemological modesty induces scientists to keep a low profile and never to forget the limited scope of their discoveries. It can be pitted against the self-confident drive of scientism which then permeated American society. As for Muir, he readily took epistemological modesty on board as a healthy obstacle to human hubris.

Muir was no doubt dedicated to scientific research and the furtherance of human knowledge. However, he was also convinced that, for all man's intelligence and resilience, the mystery of nature could not be eschewed. That is the reason why he set about defining the scope of scientific investigation. There is no question that this scope was wide but Muir felt certain that it was quite impossible to fathom the origin of life:

> There is no mystery but the mystery of harmony, no inexplicable caprice, no anomalous or equivocal expression on all the grandly inscribed mountains, although all causes that lie within reach and are readable to our limited vision are only proximate and lead on indefinitely into the impenetrable mystery of infinity. (Muir 1979, 107)

Unlike many of his fellow scientists, Muir did not feel uncomfortable about the mystery of nature since he thought that plenty of room remained for the development of human knowledge. Moreover the mystery of nature conferred to it a sacred dimension, the existence of which Muir was quick to acknowledge. One can hardly control and subdue something one does not fully comprehend, Muir reasoned. He was challenging scientism in that he was willing to countenance a measure of uncertainty when contemplating nature:

> Nature as a poet, an enthusiastic workingman, becomes more and more visible the farther and higher we go; for the mountains are fountains–beginning places, however related to sources beyond mortal ken. (Muir 1997(d), 245).

Some aspects of nature lay beyond mortal ken, so Muir believed. It is worth noting that his thirst for knowledge about the world and his acceptance of the fundamental mystery of nature went hand-in-hand. The message Muir tried to convey was that while man must be ready to study nature, he must always bear in mind that, every now and again, science may turn out to be ineffective. Hence language could prove inadequate to give a fair account of natural phenomena. As a matter of fact, part of Muir's task consisted in determining what the human mind could achieve and what it could not. There was no contradiction whatsoever between Muir's epistemological project and the modest approach to which he remained true all his life. Such a stand singled him out as a dissenting voice at a time when most people conceived of science as an instrument to establish man's dominion over nature. Muir regarded such attitudes as hubristic and vain. He did not share in the messianic enthusiasm of his contemporaries and took great pleasure in arguing that man's scientific and technical achievements should not be overstated.

Muir's epistemological modesty served to adumbrate the forging of an ecological counter-culture in the heyday of the gospel of progress. His practice of science was compatible with a proto-ecological ethos precisely because it can be pitted against men's tendency to resort to science as a tool to subjugate and exploit nature. In *The Culture of Narcissism*, Christopher Lasch opposes the promethean motives at play behind the ideology of progress to the ecological worldview that was to mature and thrive in the twentieth century and of which Muir was a forerunner:

> In psychological terms, the dream of subjugating nature is our culture's regressive solution to the problem of narcissism—regressive because it seeks to restore the primal illusion of omnipotence and refuses to accept limits on our collective self-sufficiency. [...] The science of ecology—an example of the 'exploratory' attitude toward nature, as opposed to the Faustian attitude—leaves no doubt about the inescapability of this dependence. Ecology indicates that human life is part of a larger organism and that human intervention into natural processes has far-reaching consequences that will always remain to some extent incalculable. (Lasch 1991, 244)

In the words of Christopher Lasch, Muir's brand of science appears to be exploratory rather than promethean and antagonistic towards nature. As the American century was about to begin, a dissenting voice endeavoured to substitute a proto-ecological viewpoint for the promethean spirit which characterised the age. By setting limits to the reign of science and technology in American life, Muir was acknowledging man's limited abilities in an overly optimistic era.

Works Cited

Canguilhem, George. 1994. *Etudes d'histoire et de philosophie des sciences concernant les vivants et la vie*. Paris, Librairie Philosophique J. Vrin.

Dagognet, François. 2000. *Considérations sur l'idée de nature*. Paris : Librairie Philosophique J. Vrin.

Emerson, Ralph Waldo. 2001. Nature [1836]. In *Emerson's Prose and Poetry*. Eds. Joel Porte and Saundra Morris. New York, London: W.W. Norton & Company.

Hofstadter, Richard. 1963. *Anti-Intellectualism in American Life*. New York: Vintage Books.

Jones, Maldwyn A. 1995. *The Limits of Liberty: American History 1607-1992*. 2nd Edition. Oxford: Oxford University Press, The Short Oxford History of the Modern World.

Lasch, Christopher. 1991. *The Culture of Narcissism*. New York: W.W. Norton & Company.

Muir, John. 1924. *The Life and Letters of John Muir*. Vol.2. Ed. William Frederick Bade. Boston, New York: Houghton Mifflin Company, The Riverside Press.

—. 1979. *John of the Mountains: The Unpublished Journals of John Muir* [1938]. Ed. Linnie Marsh Wolfe. Madison, WI: The University of Wisconsin Press.

—. 1994. *Steep Trails* [1918]. San Francisco: A Sierra Club Book.

—. 1997 (a). The Story of My Boyhood and Youth [1913]. In *Nature Writings*. Ed. William Cronon. New York: The Library of America.

—. 1997 (b). The Mountains of California [1894]. In *Nature Writings*. Ed. William Cronon. New York: The Library of America.

—. 1997 (c). Hetch Hetchy Valley [1912]. In *Nature Writings*. Ed. William Cronon. New York: The Library of America.

—. 1997 (d). My First Summer in the Sierra [1911]. In *Nature Writings*. Ed. William Cronon. New York: The Library of America.

—. 1998. *A Thousand-Mile Walk to the Gulf of Mexico* [1916]. Boston: Mariner Books.

—. 1999. *To Yosemite and Beyond: Writings from the Years 1863 to 1875*. Eds. Robert Engberg and Donald Wesling. Salt Lake City, UT: The University of Utah Press.

Pinchot, Gifford. 1987. *Breaking New Ground*. Covelo, California: Island Press.

—. 2004. The Fight for Conservation [1910]. In *Conservation in the Progressive Era: Classic Texts*. Seattle: University of Washington Press.

Smith, Michael L. 1987. *Pacific Visions: Scientists and the Environment 1850-1915*. New Haven: Yale University Press.

Suberchicot, Alain. 2003. *Littérature américaine et écologie*. Paris : L'Harmattan.

Thoreau, Henry David. 1973. *The Maine Woods* [1864]. Princeton, New Jersey: Princeton University Press.

—. 1992. *Walden and Resistance to Civil Government* [1854]. Ed. William Rossi. New York, London: W.W. Norton & Company.

Walls, Laura Dassow. 1995. *Seeing New Worlds: H.D. Thoreau and Nineteenth-Century Natural Science*. Madison: The University of Wisconsin Press.

Worster, Donald. 2002. *A River Running West: The Life of John Wesley Powell*. New York: Oxford University Press.

NOTES

PART I

[1] Charles Doughty, *Documents épigraphiques recueillis dans le nord de l'Arabie*, Paris: Imprimerie Nationale, 1884. A copy is kept at the National Library, one at the Bibliothèque des Langues Orientales in Paris.

[2] Jerablus (Djarábalus), north east of Aleppo.

[3] Preface to the second edition, 1888 (New York: Dover: 1979), I 31.

[4] T.E. Lawrence, Introduction to *Travels in Arabia Deserta,* 1888 (New York: Dover Publications, 1979) 25, 27.

[5] For more detail, see my book on the subject, *Ecritures du désert : Voyageurs, romanciers anglophones XIXe-XXe siècles* (Presses Universitaires de Provence, 2005).

[6] Gilles Deleuze et Félix Guattari, *Mille Plateaux*, introduction (Paris: Les Editions de Minuit, 1980).

[7] T.E. Lawrence, Introduction to *Travels in Arabia Deserta*, 19.

[8] For the history of the Ordnance Survey, see: Attorney General. House of Commons debate on 1/27/1943, § 554. *The Hansard*. Online at: <http://hansard.millbanksystems.com/commons/1943/jan/27/clause-6-additional-provisions-as-to> accessed 10/15/08

[9] "It is now proposed to make a complete and minute survey of the whole country west of the Jordan, from the extreme north to the extreme south of the Holy Land proper — 'from Dan to Beersheba' of the same nature with the Ordnance Survey of England and Wales — that is to say, not only will the natural features of the country be accurately mapped, but every town and village, every saint's tomb, every sacred tree or heap of stones, every spot, in short, to which a name is attached [...] will be faithfully plotted on our map, and its name written down in Arabic by a competent Arabic scholar [...]." See: "Palestine Exploration Fund." *Aberdeen Journal*. 11/22/1871.

[10] The testimony of Conder and Kitchener, who had first reported the attack to the WO, reached the PEF in September. It related that they faced a mob riot after Conder had tried to defend himself against the local sheikh, who had reacted violently to the investigation that was being carried out for the missing pistol. See: "The Palestine Survey Expedition." *The Times*. 09/29/1875.

[11] Royal Engineers were paid by the WO, unlike Ganneau, the French consul and explorer, who was hired and paid for by the PEF, thereby increasing the Fund's expenses.

[12] By 1877, the Sultan, who was busy conducting the Russo-Turkish War, fully realised the geopolitical stake Palestine, and more particularly Western Palestine represented, be it only because of its closeness to the Suez Canal, of which Britain

had become a minority shareholder in 1875 after Disraeli had agreed to the purchase of the Egyptian Khedive's shares. In this context, the Sultan started doubting the veracity of the archaeological and Scriptural objective of the PEF expedition, thinking that it might also be of intelligence relevance for Britain. The sultan thereby decided to impose restrictions to the firman he had delivered in December 1872, after negotiations with the FO.

[13] To back up his point, Goren quotes a letter written in 1877 by Colonel Home: "There are perhaps other reasons of a sentimental character that may perhaps be of some weight (in completing the survey) but I propose to base any recommendation on the enormous military value of a good map of the country to us." (22)

[14] For the history of the map within the WO, see WO 303 (National Archives, Kew).

[15] *Glasgow Herald*, 5/20/1876.

[16] It was Captain Burton's opinion expressed in: "Captain Burton on Turkey," *Northern Echo*, 13/3/1876.

[17] Andrew William Patrick, *The Euphrates Railway: the Shortest Route to India* (London: Effingham Wilson, 1857).

[18] See the letter by Colonel Home quoted by Goren (*Op. cit.*, 22).

[19] See, for instance: Burlington B. Wale, *The Day of Preparation; Or the Gathering of the Hosts to Armageddon: A Book for the Times* (London: E. Stock, 1893) 4.

[20] "Let this be done with the avowed intention of gradually introducing the Jew, pure and simple, who is eventually to occupy and govern this country...That which is yet to be looked for is the public recognition of the fact, together with the restoration, in whole or in part, of Jewish national life, under the protection of some one or more of the Great Powers." Charles Warren, *The Land of Promise, or Turkey's Guarantee* (1875), quoted in Franz Kobler, *The Vision Was There: A History of the British Movement for the Restoration of the Jews to Palestine* (London, 1956). Online at:
< http://www.britam.org/vision/koblerpart4.html> accessed 11/01/08

[21] John Wilson published a seminal work entitled *Our Israelitish Origin: Lectures* (London, 1844).

[22] On British Israelism, see: Richard Simpson, 'The political influence of the British-Israel Movement in the Nineteenth Century", 2002. Online at:
<http://www.originofnations.org/books,%20papers/MA_dissertation_BI.pdf>accessed 11/01/08

[23] In the first issue of *The Banner of Israel*, appeared the following statement in the introduction: "We are living in solemn times! The Eastern Question is before us in terrible proportions, insoluble apparently, except by the awful arbitrament of the sword! 'The Mighty Earthquake of the Nations' which is to rend all Gentile polities to their very foundations, is at the doors! The time for Israel's 'Return', with Judah, to their own inheritance, after the acquisition by the former of their promised 'Gate', is *the* event we anticipate as close at hand." See: "Our New Year," *The Banner of Israel*, N°1 (3 January, 1877), London, 1.

[24] "The Exploration of Palestine Completed," *The Banner of Israel*, N°46, 14 November, 1877, 376.

[25] In this letter, Conder also added that this new colonisation scheme could only be successful if the building of the railway from Haifa to Damascus was completed. He considered this possible, especially as the port of Haifa was already being envisaged as the terminus for the Euphrates Valley Railway. Clearly, the Western Survey map was at the origin of numerous debates, Biblical, geographical and politico-imperial.

[26] Hughenden Papers, Dep. Hughenden 138/2, ff. 59-62, Oliphant to Corry, 26 May [1879?]

[27] Laurence Oliphant, *The Land of Gilead: With Excursions in the Lebanon* (London & Edinburgh: W. Blackwood & Sons, 1880) xxxii-xxxiii. He wrote: "It is somewhat unfortunate that so important a political and strategical question as the future of Palestine should be inseparably connected in the public mind with a favourite religious theory. I would avail myself of this opportunity of observing that, so far as my own efforts are concerned, they are based upon considerations which have no connection whatever with any popular religious theory upon the subject."

[28] Precisely, the Jewish weekly approved of this project, deeming it replete with "very practicability and non-sentimentality" ("Mr Laurence Oliphant's Scheme for the Colonisation of Palestine," *The Jewish Chronicle*, 9 January, 1880, in Laurence Oliphant, *Op. cit.*, 526). But in this review published on 9 January, 1880, the author of the article "Mr Laurence Oliphant's Scheme for the Colonisation of Palestine" praised the scheme, highlighting that it could both be interpreted from the point of view of eschatology and of practical politics. He wrote: "It is oppression, and not prosperity, which will lead us [Jews] back to our proper place in the Holy Land. It cannot be denied that at no period of our modern history have there been so many forces at work which tend directly to the Great Restoration. Signs and portents abound, and the air is thick with rumours. Can these be the precursors of the Event, or are they but evidence of the restless spirit of advanced civilisation?" This hovering and indeterminacy between the political and the eschatological, between "the political center-stage and the religious fringe" (Eitan Bar-Yosef, 29), stemmed from the political instability in European and Asiatic Turkey and was actually a characteristic of the era. Another example could be Cazalet's colonisation scheme in Palestine, which would encourage the settlement of Jews along the Euphrates Railway. Cazalet, a merchant and an industrialist, saw this scheme as being of British responsibility, especially towards Syria. Cazalet also thought that this scheme would extend British (imperial) influence. He also conceived the restoration of the Jews as being part of "England's mission," to take up Gladstone's phrase. See Edward Cazalet, *The Eastern Question: An Address to Working Men, with map showing the projected line of the Euphrates Valley Railway* (London: Edward Stanford, 1878) 42.

[29] See *A Map of Palestine from Surveys Conducted for the Committee of the Palestine Exploration Fund and Other Sources*, London, 1890.

[30] Eitan Bar-Yosef writes: "Without its eschatological backbone, resting, as it

were, on a purely imperial basis, the strategic logic behind the Jewish colonization of Palestine seemed flawed and insufficient. The imperial vocabulary would perhaps eclipse the millenarian madness, render it respectable, but the eclipse was never full; and the Christian Zionists themselves were the first to sense this." (34)

[31] The map (now in the collections of the British Museum) is called after its 17th century owner, Sir Robert Cotton (1571-1631). It is a hand-coloured map on vellum carrying Elizabethan emblems though it admittedly dates back to a previous epoch. See "A 16th century map of the British Isles", Royal Geographical Society, *The Geographical Journal*, Vol. 34, n.4, October 1909, 421-423.

PART II

[32] See famous French ethnologist Glowczewski's book on the dreamtime: *Du rêve à la loi chez les Aborigènes. Mythes, rites et organisation sociale en Australie* (1999).

[33] For a thorough discussion of this question, see Moira Simpson's *Making Representations: Museums in the Post-Colonial Era*, London: Routledge, 1996.

[34] Turnbull subsequently clarified his position by pointing out that though "museums acquired skeletons from grave-robbers, doctors, and from aborigines (sic) killed in frontier massacres," the idea that they actually incited murder "can not be sustained" (Rennie 1998, n.p.). But while he exculpates the museums themselves, he does not deny that murder took place for the purpose of providing museums and the scientific community more generally with the Aboriginal remains they were so avid to acquire.

[35] An extensive body of evidence documenting the links that existed between anthropological research and colonial government in Australia can be found both in Richard Glover's "Scientific Racism and the Australian Aboriginal (1865–1915)", Jan Kociumbas (ed), *Maps, Dreams, History* and in John Cove's *What the Bones Say*.

[36] The evolutionary paradigm contained in fact, a double message since, as Coombes argues, as well as confirming the biological superiority of the white race, it also emphasised "the inevitability and indispensability of the existing social order and its attendant inequalities" (1994, 121). Ethnographic displays at International Exhibitions were motivated by the same two-fold ideological aims. Through them, observes Robert Rydell, "anthropologists (…) sought to educate the public about the applicability of social Darwinian insights to social struggle at home and imperial expansion abroad." (1999, 136)

[37] The details of this story have all been gleaned from Helen MacDonald's *Human Remains* (2005) and Robert K. Hitchcock's "Repatriation, indigenous peoples, and development lessons from Africa, North America, and Australia" (2002).

[38] Such a bloodless acceptance of the priority of scientific knowledge over human values did not, it seems, come without a cost since shortly after Ishi's death, Kroeber suffered from severe depression and abandoned his anthropological career

for several years during which he underwent psychoanalysis (Scheper-Hughes 2002, 361).

[39] At a hearing on the repatriation of Ishi's remains in 1999, Scheper-Hughes, as a member of Berkeley's Department of Anthropology, read a statement in which she apologized for her department's role in "the final betrayal of Ishi, a man who had already lost all that was dear to him at the hands of Western colonisers" (2002, 363).

[40] Obtaining a second degree from Europe was common practice in the early 19th century for American MD's in order to achieve higher standards (Gillet 1987, 22).

[41] After 1832, the US army accepted physicians only after they had passed an examination; as late as 1860, 50% of the applicants who were rejected because they failed the exam went on to practice medicine as civilians (Gillet 1987, 23). The army surgeons were therefore professionals, the best qualified in the field. The fact that Morton relied on the doctors for his collections illustrated the soundness of his scientific approach.

PART III

[42] My definition of the term 'Arcadian' is that of a compromise, a blissful union of city and countryside, of urban and rural values. The term thus includes but transcends the notion of agrarianism, promoting a predominantly agrarian economy, to denote, as stressed by Peter J. Schmitt, a philosophical idea emphasizing the spiritual impact nature is said to have upon mankind. Cf Schmitt, 1990, *Introduction*.

[43] This was the recurrent criticism raised by the Federalists and their leader, Alexander Hamilton, the author of the *Report on the Subject of Manufactures*, presented to Congress in 1791 and advocating the need for the country's rapid industrialization. Cf. Peterson 1970, 581.

[44] Cf. *The Correspondence of Jefferson and Du Pont de Nemours, with an introduction on Jefferson and the Physiocrats*, Gilbert Chinard, 1931 (Paris: Belles Lettres). See also the letter, dated Feb. 1st, 1804, sent by Jefferson to Jean-Baptiste Say, in which he discusses the latter's *Political Economy* and the merits for America to specialize in agriculture (Jefferson 1984, 1143-1144).

[45] Leo Marx, in his famous *The Machine in the Garden*, describes Jefferson's irenic vision and his quest for a "middle landscape." Cf. pp. 116-150.

[46] Cf. *Technological Utopianism in American Culture*, Howard P. Segal. (Syracuse: Syracuse University Press, 2005). See Jefferson's letter of February the 8th, 1805, to the Comte de Volney, for his ideas for New Orleans (Jefferson 1984, 1157).

[47] Another revealing paradox, pointed out by Peter J. Schmitt in his book (1990), is that Turner expressed his famous Frontier theory, singing the praise of agrarian America much in the way Jefferson could have done, in the context of a World Fair celebrating the technological and urban values of America.

[48] A contemporary student of utopian communities in America explained the failure of so many of them by their obsession with land and farming and their almost complete oblivion of manufacturing (Noyes 1966, 19-20).

[49] This description of the three states of development separating the "savages" of the western wilds, the "barbarians" of the Frontier and the "civilized" inhabitants of the eastern coast would be adopted in a strikingly similar fashion by Turner in his 1893 speech, with Turner also lauding, as seen previously, the "barbarians" as, implicitly, the best members of society.

[50] Among the illustrations of Jefferson's interests for science's practical uses can be mentioned, outside the realm of agriculture, his interest for Robert Fulton's invention of a submarine and of torpedoes that would have allowed to efficiently protect the country from military assaults (Bedini 1990, 385).

[51] *Thomas Jefferson, Apostle of Americanism*, Gilbert Chinard (New York: Overseas Editions, 1939).

[52] For examples of opposite viewpoints on the issue, see Merk, Frederick. *Manifest Destiny and Mission in American History* (Cambridge, Mass., Harvard University Press, 1995); LaFeber, Walter. *The New Empire: An Interpretation of American Expansion, 1860-1898.* (Ithaca, Cornell University Press, 1963); Hietala, Thomas. *Manifest Design: Anxious Aggrandizement in Late Jacksonian America.* (Ithaca, Cornell University Press, 1985).

[53] Benjamin Franklin, *A Dissertation on Liberty and Necessity, Pleasure and Pain* (1725); The quote is borrowed from Commager (1977), *The Empire of Reason*, 2.

[54] The phrase is the title of one of Hume's *Political Essays*, which is also available in the electronic Library of Economics and Liberty, at the following site: http://www.econlib.org/library/LFBooks/Hume/hmMPL3.htlm

[55] The quote is from Barry Cough, *Distant Dominion: Britain and the Northwest Coast of North America*, Vancouver, 1980 and is borrowed from Nacouzi, Salwa. "Thomas Jefferson et les raisons d'une expédition : exploration, expansion, expansionnisme" in Caron, Nathalie & Naomie Wulf, *The Lewis and Clark Expedition* (Paris: Editions du temps, 2005).

[56] Jefferson to Monroe, Nov. 24, 1801. Quote borrowed from Tucker, Robert W. & Hendrickson, David C. *Empire of Liberty. The Statecraft of Thomas Jefferson* (New York : Oxford University Press, 1990) 160-161. A few years later, in a private letter to Madison, Jefferson also suggested that Cuba was the southernmost natural border of the United States. (Nacouzi, *op.cit.*, 9-20).

[57] Jefferson, Letter to Alexander von Humboldt, 1813, quoted by Whitaker, Arthur P. *The Western Hemisphere Idea* (New York: Ithaca, 1954) 29.

[58] Secretary of State Seward stated in 1859 that Cuba would come to the U.S. "by means of constant gravitation," and Representative Godlove Orth explained in 1870 that Cuba "must inevitably gravitate towards us." Both quotes appear in Weinberg, *op. cit.*, 234.

[59] The quote is an extract from the *United States Democratic Review* (1858), and is borrowed from Weinberg (1963, 236).

[60] "It may be taken for a certainty that not a foot of land will ever be taken from the Indians without their consent," Jefferson wrote in 1786, even though the future would soon contradict him (Weinberg 1963, 72).

[61] For an analysis of *Machtpolitik*, see Friedrich Meinecke's classic study, *Machiavellism, the Doctrine of Raison d'Etat and Its Place in Modern History*, as well as Aron, *Paix et guerre entre les nations*, 575-582.

[62] The following analysis of the influence of German thought on John Burgess is based upon Charles Robson's unpublished thesis: *The Influence of German Thought on Political Theory in the United States in the Nineteenth Century* (Ph.D. History department, University of North Carolina, 1930) 314-337.

PART IV

[63] Paul Semonin traces the paradigm of dominance that dinosaur images have been used for in "Empire and Extinction: The Dinosaur as a Metaphor for Dominance in Prehistoric Nature," *Leonardo*. Vol. 30, no. 3 (1997), 171-182.

[64] Donna Haraway, *Primate Visions: Gender, Race and Nature in the World of Modern Science* (New York: Routledge, 1989). See chapter "Teddy Bear Patriarchy Taxidermy in the Garden of Eden, New York City, 1908-1936." Mark Meigs, "À la recherche d'un passé visible: regards publics et privés dans les collections de la région de Philadelphie, 1890-1930" *Publics & Musées*, 16, Montréal, (juillet et décembre 1999) 66-68. The memorializing of the adventure took place not only in the diorama itself, but in a whole series of semi-scientific journals proliferated by museums during this period, that described the specimen collecting trips not in terms of science but in terms of exciting tales.

[65] See, Richard C. Ryder, "Dusting off America's First Dinosaur," *American Heritage*, March 1988, Vol. 39, issue 2.

[66] Uintatherium robustum, 1872.
www.ansp.org/museum/leidy/paleo/uintatherium.php

[67] Leonard Warren, *Joseph Leidy: The Last Man Who Knew Everything* (New Haven: Yale University Press, 1998), 175, describes Leidy's place in the scientific world. He is credited with 230 publications on paleontology. His four major books on paleontology are: *The Ancient Fauna of Nebraska*, Washington, D. C. (Smithsonian Contributions to Knowledge, 1853 (Accepted for publication in 1852); *Cretaceous Reptiles of the United States*, Washington, D. C. (Smithsonian Contributions to Knowledge, 1865); *On the Extinct Mammalia of Dakota and Nebraska, Philadelphia* (Journal of the Academy of Natural Sciences,1869); and *Extinct Vertebrate Fauna of the Western Territories*, Washington, D. C. (Government Printing office, 1873).

[68] Warren, *Joseph Leidy*, 226.

[69] Quoted in Warren, *Joseph Leidy*, 228. Warren quoted a letter to from Leidy to Rev. Henry C. McCook, February 9, 1879, in which Leidy quotes John Fiske's *Outlines to Cosmic Philosophy*, of 1874. McCook, a member of the Academy of

Natural Sciences, had asked Leidy specifically to state his beliefs for the benefit of his congregation.

[70] See, Peter J. Bowler, "Edward Drinker Cope and the Changing Sructure of Evolutionary Theory," in *Isis*, Vol. 68., No. 2 (June 1977), 249-265; Edward J. Pfeifer, "The Genesis of American Neo-Lamarckism," in *Isis*, Vol. 56, No. 2 (Summer, 1965), 156-167.

[71] Edward Drinker Cope, "The Origins of Genera," *Proceedings of the Acacemy of Natural Sciences of Philadelphia*, vol. 20, 1868, 242-300 (specific quotes taken from page 243 and the conclusion on page 300.

[72] Edward Drinker Cope, "Laelaps Cope," *Proceedings of the Academy of Natural Sciences of Philadelphia*, vol. 20, 1868, 241.

[73] The story has been often told in scientific but also popular forms. See Mark Jaffe, *The Gilded Dinosaur: The Fossil War Between E. D. Cope and O. C. Marsh and the Rise of American Science* (New York: Crown, 2000); James Penick. "Professor Cope vs. Professor Marsh: A bitter feud among the bones," *American Heritage*, August 1971, Vol. XXII, no 5; "Bone Wars: the Cope-Marsh Rivalry" ww.ansp.org/museum/leidy/paleo/bone_wars.php

[74] Paleontologists had relied on railroad engineers and military men for much of the field work involved in early American paleontology, by the end of the nineteenth century they had come to rely more on oil men.

[75] For Carnegie and Dinosaurs see Paul Semonin, "Empire and Extinction: The Dinosaur as a Metaphor for Dominance in Prehistoric Nature," in *Leonardo*, Vol. 30, no. 3, 1997, 171-181. MIT Press. Carnegie published his essay "Wealth," in 1889, a good decade before he embarked on dinosaurs.

[76] See Michael Crichton, *Jurassic Park* (New York: Alfred Knopf, 1990); *The Lost World* (New York: Alfred Knopf, 1995) a sequel to the first novel. Crichton also wrote an introduction to an edition of Sir Arthur Conan Doyle's *The Lost World* (New York: Modern Library, 2003). The movie, "Jurassic Park" appeared in 1993 and the movie sequel appeared in 1997. A second sequel, "Jurassic Park III," appeared in 2001. A fourth installment was abandoned.

[77] From the Latin, *evolvere,* to unfold; from *volvere,* to roll; also *evolutio,* an unrolling.

[78] See Martin Rudwick's classic study, *The Meaning of Fossils* (1976).

[79] In *The Non-Darwinian Revolution*, Peter Bowler argues that the Darwinian theory was assimilated in a bastardised form that exaggerated the inevitability of progress.

[80] There has been much debate among historians of science as to whether Buffon can be considered a transmutationist or not; it is at least clear that the narrative outlined in *Epoques* does not require such an outlook.

[81] The escalator analogy was proposed by C.C. Gillispie in his essay "Lamarck and Darwin in the History of Science," in Glass, Temkin and Straus (eds.), *Forerunners of Darwin, 1745-1859.*

PART V

[82]*Lettres Edifiantes et Curieuses écrites des Missions Etrangères,* nouvelle édition (Paris: Mérigot Le Jeune, 1781) « Mémoires d'Amérique », VI: j-ij. *The Jesuit Relations* (41 vol.) were first published in Paris from 1632 till 1672. The collection of the *Edifying and Curious Letters* (34 vol.) was first published in Paris from 1703 till 1776, followed by several reprints until the 1850s. Starting in 1822 the *Nouvelles des Missions,* then in 1825 *The Annals of the Association of the Propagation of Faith* until the eve of WW 2, in which Peter-John De Smet published several letters between 1840 and 1865. The *Annals* were widely circulated: 10,000 copies until 1830 and 15,000 copies onwards. Various Spanish, Italian and German translations were simultaneously published. Starting in 1868, the Association of the Propagation of Faith also published a weekly bulletin, *Les Missions Catholiques,* widely translated in Western Europe.

[83] See also Francis Parkman. *The Jesuits in North America in the Seventeenth Century* (Williamstown, Mass: Corner House, 1980); as well as G. Mengarini. The Rocky Mountains: The Memoirs of Father Gregory Mengarini, in *Woodstock Letters* 18, 1889, 32.

[84] Campanella wrote to Galileo from his own prison: "After your message, Ô Galileo, the whole Science has to be renewed." Quoted by Georges, Jean. *Voyages en Utopie* (Paris: Gallimard, 1994) 52.

[85] "A M. Rollier, avocat à Opdorp, près de Termonde. Rivière Saint-Ignace, 10 septembre 1841." Washington State University Libraries, Pullman, Washington State 99164-5610 / Holland Library / Manuscripts, Archives & Special Collections, Cage 537, 1, 11, 410910.

[86] Born and educated in Switzerland, Louis Agassiz arrived in New England in 1846. There he fast attained the status of a celebrity. The fledging American scientific community looked up to him so much so that he was appointed professor of zoology and botany at Harvard University only a year after his arrival in the United States. In the 1850s Agassiz succeeded in building a wide network in American scientific circles. He took control of key institutions such as the American Association for the Advancement of Science and the Smithsonian Institution. His influence was then at its utmost. Matters came to head in the 1860s.

[87] Elizabeth B. Keeney, *The Botanizers: Amateur Scientists in Nineteenth-Century America* (Chapel Hill: The University of North Carolina Press, 1992) 23.

[88] Once in Florida Muir was unfortunately struck down by malarial fever and had to cancel his plan. However his fascination with Humboldt's venture was such that he eventually sailed to Brazil in 1911, at seventy-three years of age.

BIBLIOGRAPHY

Adair, Douglas. 1974. That Politics May Be Reduced to a Science. David Hume, James Madison and the 10th Federalist. In *Fame and the Founding Fathers*. New York: Norton and Co.

Adams, Brooks. 1900. *America's Economic Supremacy*. New York: MacMillan.

—. 1943. *The Law of Civilization and Decay*. New York: Vintage Books.

Adams, Charles Francis. 1874-77. *Memoirs of John Quincy Adams*. 12 vols, Philadelphia: J.B. Lippincott.

Adas, Michael. 1989. *Machines as the Measure of Men: Science, Technology and Ideologies of Western Dominance*. Ithaca and London: Cornell U. P.

Agassi, Joseph. 2008. *Science and Its History: A Reassessment of the Historiography of Science*. Boston Studies in the Philosophy of Science, Vol. 253. Spring.

Anderson, Warwick. 2006. *The Cultivation of Whiteness: Science, Health, and Racial Destiny in Australia*. Durham, NC: Duke University Press.

Andrews, J. H. 1974. *History in the Ordnance Map: An Introduction for Irish Readers*. Dublin: Ordnance Survey.

—. 1975. *A Paper Landscape: The O. S. in Nineteenth-Century Ireland*. Oxford: Clarendon.

—. 1997. *Shapes of Ireland: Maps and Their Makers 1564-1839*. Dublin: Geography.

Aron, Raymond. 1984. *Paix et guerre entre les nations*. Paris: Calman-Lévy.

Asad,Talal. 1973. *Anthropology and the Colonial Encounter*. Amherst (NY): Humanity Books.

Ballantyne, Tony. 2004. *Science, Empire and the European Exploration of the Pacific*. Aldershot: Ashgate Variorum.

Barnes, Trevor and James S. Duncan. 1992. *Writing Worlds. Discourse, Text and Metaphor in the Representation of Landscape*. Londres : Routledge.

Bates, Daisy. 1972. *The Passing of the Aborigines: A Lifetime Spent among the Natives of Australia* [1938]. London: Panther.

Bedini, Silvio A. 1990. *Thomas Jefferson: Statesman of Science*. New York: Macmillan Publishing Company.

Bibliography

Bentham, Jeremy. 1776. *A Fragment On Government.* http://onlinebooks.library.upenn.edu, Jan 28, 2009.

—. 1812. *The Panopticon versus New South Wales, Two Letters to Lord Pelham.* London: R. Baldwin.

Boller, Paul. 1970. *American Thought in Transition: The Impact of Evolutionary Naturalism, 1865-1900.* Chicago: Rand McNally.

Bourguet, Marie-Noëlle & Christophe Bonneuil. 1999. De l'Inventaire du monde à la mise en valeur du globe: Botanique et colonisation. In *Revue Française d'Histoire d'Outre-mer* 86 : 322-323.

Bowler, Peter. 1988. *The Non-Darwinian Revolution. Reinterpreting a Historical Myth.* Baltimore and London: John Hopkins University Press.

Boyd, Julian. 1968. Thomas Jefferson's "Empire of Liberty." In Peterson, Merrill. *Thomas Jefferson.* New York: Hill and Wang.

Buckley, Arabella. 1882. *Winners in Life's Race; or, the Great Backboned Family.* London: Macmillan.

Buffon, Georges Leclerc, Comte de. 1988. *Les Epoques de la nature* [1780]. Paris : Editions du Muséum national d'Histoire naturelle. (orig. pub..)

Brantlinger, Patrick. 2003. *Dark Vanishings: Discourse on the Extinction of Primitive Races, 1800-1930.* Ithaca (N.Y.) & London: Cornell U.P.

Brown, Malcolm Ed. 1991. *The Letters of T.E. Lawrence.* London: J.M. Dent and Sons Ltd. Oxford: Oxford University Press.

Canguilhem, George. 1994. *Etudes d'histoire et de philosophie des sciences concernant les vivants et la vie.* Paris, librairie philosophique J. Vrin.

Carter, Paul. 1988. *The Road to Botany Bay. An Exploration of Landscape and History.* New York: Alfred A. Knopf.

Casey, Edward S. 2002. *Representing Place. Landscape Painting and Maps.* Minneapolis, London: University of Minnesota Press.

Cazalet, Edward. 1878. *The Eastern Question: An Address to Working Men, with Map Showing the Projected Line of the Euphrates Valley Railway.* London: Edward Stanford.

Chambers, Robert. 1994. *The Vestiges of the Natural History of Creation, and Other Evolutionary Writings* [1844]. Chicago: University of Chicago Press. (Facsimile edition)

Chesterman, John & Brian Galligan. 1997. *Citizens Without Rights: Aborigines and Australian Citizenship.* Cambridge: Cambridge U. P.

Chittenden, Hiram Martin & Alfred Talbot Richardson. 1905. *Life, Letters and Travels of Father Pierre-Jean De Smet, s.j., 1801-1873.* New York: Harper.

Clarke, Marcus. 1877. *The Future Australian Race*. Melbourne: A. H. Massima.

Commager, Henry Steele. 1977. *The Empire of Reason*. New York: Oxford University Press.

Conder, Claude Reigner. 1879. *Tent Work in Palestine: A Record of Discovery and Adventure*. Vol. 2. London: Richard Bentley & Son.

Condorcet, Jean-Antoine-Nicolas Caritat. 1988. *Esquisse d'un Tableau historique des progrès de l'esprit humain, suivi de Fragment sur l'Atlantide* [1795]. Paris: Flammarion.

Coombes, Annie. 1994. *Reinventing Africa: Museums, Material Culture and Popular Imagination*. New Haven & London: Yale U.P.

Cove, John J. 1995. *What the Bones Say: Tasmanian Aborigines, Science and Domination*. Ottawa: Carleton U.P.

Dagognet, François. 2000. *Considérations sur l'idée de nature*. Paris : Librairie philosophique J. Vrin.

Deleuze, Gilles. 1992. *The Fold: Leibniz and the Baroque*. University of Minnesota Press.

Deleuze, Gilles & Félix Guattari. 2004. *A Thousand Plateaus*. Trans. Brian Massumi. London and New York: Continuum. Vol. 2 of *Capitalism and Schizophrenia*. 2 vols. 1972-1980. Trans. of *Mille Plateaux*. Paris: Les Editions de Minuit.

Delmas, Catherine. 2005. *Ecritures du désert : Voyageurs, romanciers anglophones XIXe-XXe siècles*. Aix en Provence: Presses universitaires de Provence.

De Smet, Pierre-Jean, s.j. 1875. *Lettres Choisies, 1849-1857*. Bruxelles: Closson et Cie.

—. 1859. *Western Missions and Missionaries: A Series of Letters*. New York: P.J. Kenedy.

Doherty, Gillian M. 2004. *The Irish O.S., History, Culture and Memory*. Belfast: Four Courts Press.

Doughty, Charles. 1979. *Travels in Arabia Deserta* [1888]. 2 vols. New York: Dover.

—. 1884. *Documents épigraphiques recueillis dans le nord de l'Arabie*. Paris : Imprimerie Nationale.

Drouin, Jean-Marc. 2008. *L'herbier des philosophes*. Paris: Seuil.

Emerson, Ralph Waldo. 2001. Nature. In *Emerson's Prose and Poetry* [1836]. Joel Porte and Saundra Morris eds. New York, London: W.W. Norton & Company.

Fabian, Johannes. 1983. *Time and the Other: How Anthropology Makes its Object*. New York: Columbia U.P.

Farrand, Max. 1937. *The Records of the Federal Convention of 1787.* New Haven: Yale University Press.

Foster, John Wilson. 1997. Encountering Tradition. In *Nature in Ireland : a Scientific and Cultural History.* Dublin: Lilliput Press.

Foucault, Michel. 1972. *The Archaeology of Knowledge.* Trans. A. M. Sheridan Smilth. New York: Pantheon.

—. 2002. *The Order of Things : An Archeology of the Human Sciences.* Routledge. Trans of *Les mots et les choses.* Paris: Gallimard.

Friel, Brian. 1981. *Translation.* London: Faber and Faber.

Frodsham, G. H. 1915. *A Bishop's Pleasaunce.* London: Smith Elder.

Fülöp-Miller, René. 1933. *Les Jésuites et le Secret de leur Puissance. Histoire de la Compagnie de Jésus.* Paris: Plon.

Gascoigne, John. 1998. *Science in the Service of Empire. Joseph Banks, the British State and the Uses of Science in the Age of Revolution.* Cambridge UP.

Gilbert, Felix. 1961. *To the Farewell Address. Ideas of Early American Foreign Policy.* Princeton, New Jersey: Princeton University Press.

Gillett, Mary C. 1987. *The Army Medical Department, 1818-1865.* Washington D.C.: Center of Military History, U.S. Army.

Gillispie, Charles Coulston. 1959. Lamarck and Darwin in the History of Science. In *Forerunners of Darwin, 1745-1859,* eds. Glass, Temkin and Straus. Baltimore: John Hopkins Press.

Glowczewski, Barbara. 1991. *Du rêve à la loi chez les Aborigènes. Mythes, rites et organisation sociale en Australie.* Paris: PUF/Ethnologies.

Gould, Stephen Jay. 1981. *The Mismeasure of Man.* New York & London: W. W. Norton & Co.

Graebner, Norman. 1985. *Foundations of American Foreign Policy.* Wilmington, Delaware: Scholarly Resources Inc.

Graham B. J. & L. J. Proudfoot. 1993. *An Historical Geography of Ireland.* London, San Diego: Academic Press.

Greene, John. 1984. *American Science in the Age of Jefferson.* Ames, Iowa: Iowa State Press.

Hallam, Elizabeth & Brian Street (eds). 2000. *Cultural Encounters: Representing Otherness.* London & NY: Routledge.

Harding Sandra (ed). 1993. *The 'Racial' Economy of Science: Toward a Democratic Future.* Bloomington (Ind): Indiana U.P.

Harley, Brian. 1992. Deconstructing the Map. In *Writing Worlds: Discourse, Text and Metaphor in the Representation of Landscape.* Ed. Barnes, Trevor & James Duncan. London, New York: Routledge.

Haraway, Donna. 1989. *Primate Visions: Gender, Race and Nature in the World of Modern Science*. New York: Routledge.

Haubert, Maxime. 1967. *La Vie Quotidienne au Paraguay sous les jésuites*, Paris: Hachette.

Helgerson, Richard. 2001. The Folly of Maps and Modernity. In *Literature, Mapping, and the Politics of Space in Early Modern Britain*. Ed. Andrew Gordon & Bernhard Klein. Cambridge: Cambridge University Press.

Hietala, Thomas. 1985. *Manifest Design: Anxious Aggrandizement in Late Jacksonian America*. Ithaca: Cornell University Press.

Hindle, Brooke. 1974. *The Pursuit of Science in Revolutionary America, 1735-1789*. New York: Norton & Company.

Hofstadter, Richard. 1959. *Social Darwinism in American Thought*. Boston: The Beacon Press.

—. 1963. *Anti-Intellectualism in American Life*. New York: Vintage Books.

Hogarth, David George. 1904. *The Penetration of Arabia*. London: Lawrence and Bullen.

—. 1929. *The Life of Charles M. Doughty*. New York: Doubleday.

Howes, Marjorie. 2000. Goodbye Ireland I'm Going to Gort: Geography, Scale, and Narrating the Nation. In *Semicolonial Joyce*. Ed. Attridge & Howes, University of York: Rutgers University.

Hughes, Robert. 1996. *The Fatal Shore: A History of the Transportation of Convicts to Australia, 1787–1868*. London: Harvill Press.

Hume, David. 1953. *Political Essays* [1752]. New York: Charles Hendel.

Hunt Jackson, Helen. 1964. *A Century of Dishonor, a sketch of the United States Governments Dealings with some of the Indians Tribes*. Minneapolis: Minnesota, Ross & Haines.

Jackson John P. Jr. & Nadine M. Weidman. 2004. *Race, Racism and Science: Social Impact and Interaction*. Santa Barbara (Cal.): ABC-CLIO.

Jacyna, Stephen. 2006. Medicine in Transformation 1800-1849. In *The Western Medical Tradition*. Eds. W. F. Bynum & al. New York (N.Y.): Cambridge University Press.

Jaffe, Mark. 2000. *The Gilded Dinosaur: The Fossil War Between E. D. Cope and O. C. Marsh and the Rise of American Science*. New York: Crown.

Jean, Georges. 1994. *Voyages en Utopie*. Paris: Gallimard.

Jefferson, Thomas. 1984. *Writings*. New York: Literary Classics of the United States.

—. 1903. *The Writings of Thomas Jefferson*. Vol. 12.Washington: The Thomas Jefferson Memorial Foundation of the United States.

Jones, Maldwyn A. 1995. *The Limits of Liberty: American History 1607-1992 2nd Edition*. Oxford: Oxford University Press.

Kain, J. P. & E.Baigent. 1984. *The Cadastral Map in the Service of the State, a History of Property Mapping*. Chicago: The University of Chicago Press.

Knipe, Henry R. 1905. *Nebula to Man*. London: J.M. Dent & Co.

—. 1912. *Evolution in the Past*. London: Herbert and Daniel.

Kobler, Franz. 1956. *The Vision Was There: A History of the British Movement for the Restoration of the Jews to Palestine*. London. < http://www.britam.org/vision/koblerpart4.html> accessed 11/01/08

Kociumbas, Jan (ed). 1998. *Maps, Dreams, History: Race and Representation in Australia*. Sydney: Braxus Publishing.

—. 2004. Genocide and Modernity in Colonial Australia, 1788-1850. In *Genocide and Settler Society* ed. A. Dirk Moses. New York: Berghahn Books.

Lacouture, Jean. 1991-1992. *Jésuites*. Paris: Seuil.

LaFeber, Walter. 1963. *The New Empire: An Interpretation of American Expansion, 1860-1898*. Ithaca: Cornell University Press.

Lasch, Christopher. 1991. *The Culture of Narcissism*. New York: W.W. Norton & Company.

Lawrence, Thomas Edward. 1979. Introduction to *Travels in Arabia Deserta* [1888], 2 vols. New York: Dover.

Le Squère, Roseline. 2006. Analyse des perceptions, usages et fonctions des toponymes actuels des territoires ruraux et urbains de Bretagne. In *Noms propres, dynamiques identitaires et socio-linguistiques*. Ed. Francis Manzano. Rennes: Presses Universitaires de Rennes.

Lorimer, Douglas. 1996. Race, science and culture : historical continuities and discontinuities, 1850-1914. In *The Victorians and Race*. Ed. Shearer West. Aldershot: Scholar Press.

Love, Eric. 2007. *Race Over Empire : Racism and U.S. Imperialism, 1865-1900*. Chapel Hill: University of North Carolina Press.

MacDonald, Helen. 2006. *Human Remains: Dissection and its Histories*. New Haven & London: Yale U.P.

McDougall, Walter. 1997. *Promised Land, Crusader State. The American Encounter with the World Since 1776*. New York: Houghton Mifflin.

MacLeod, Roy, ed. 2000. Nature and Empire: Science and the Colonial Enterprise. *Osiris: A Research Journal Devoted to the History of Science and its Cultural Influences*. Second Series. Vol.15.

Mahan, Alfred. 1897. *The Influence of Sea Power upon History*. Boston: Little, Brown and Co.

—. 1902. *Retrospect and Prospect*. Boston: Little, Brown and Co.

—. 1907. *Some Neglected Aspects of War*. Boston: Little, Brown and Co.

—. 1912. *Armaments and Arbitration or the Place of Force in the International Relations of States*. New York: Harper and Brothers.

—. 1915. *The Interest of America in International Conditions*. Boston: Little, Brown and Co.

Maier, Charles. 2007. *Among Empires: American Ascendancy and Its Predecessors*. Cambridge, Mass.: Harvard University Press.

Malouf, David. 1994. *Remembering Babylon*. London, Sydney: Vintage.

Meinecke, Friedrich. 1997. *Machiavellism, the Doctrine of Raison d'Etat and Its Place in Modern History*. Philadelphia: Transaction Books.

Menand, Louis. 2001. Morton, Agassiz, and the Origins of Scientific Racism in the United States. *The Journal of Blacks in Higher Education*.

Merk, Frederick. 1995. *Manifest Destiny and Mission in American History*. Cambridge, Mass.: Harvard University Press.

Minois, Georges. 1990. *L'Eglise et la Science. Histoire d'un malentendu*. Paris: Fayard.

Monmonier, Mark. 1991. *How to Lie with Maps*. Chicago: University of Chicago Press.

Morton, Samuel George. 1844. *Crania Aegyptiaca. Observation from Aegyptian Ethnography Derived from Anatomy, History, and the Monuments*. Philadelphia: Penington.

Morton, Samuel George. 1839. *Crania Americana; or , a Comparative View of the Skulls of Various Aboriginal Nations of North and South America. To Which Is Prefixed an Essay on the Varieties of the Human Species*. 1st ed. Philadelphia: J. Dobson.

Moscrop, John James. 2000. *Measuring Jerusalem: The Palestine Exploration Fund and British Interests in the Holy Land*. London & New York: Leicester University Press.

Muir, John. 1924. *The Life and Letters of John Muir*. Vol.2. Ed. William Frederick Badè. Boston, New York: Houghton Mifflin Company, The Riverside Press.

—. 1979. *John of the Mountains: The Unpublished Journals of John Muir* [1938]. Ed. Linnie Marsh Wolfe. Madison, WI: The University of Wisconsin Press.

—. 1994. *Steep Trails* [1918]. San Francisco: A Sierra Club Book.

—. 1997 (a). The Story of my Boyhood and Youth [1913]. In *Nature Writings*. Ed. William Cronon. New York: The Library of America.

—. 1997 (b). The Mountains of California [1894]. In *Nature Writings.* . Ed. William Cronon. New York: The Library of America.

—. 1997 (c). Hetch Hetchy Valley [1912]. In *Nature Writings.* Ed. William Cronon. New York: The Library of America.

—. 1997 (d). My First Summer in the Sierra [1911]. In *Nature Writings.* Ed. William Cronon. New York: The Library of America.

—. 1998. *A Thousand-Mile Walk to the Gulf of Mexico* [1916]. Boston: Mariner Books.

—. 1999. *To Yosemite and Beyond: Writings from the Years 1863 to 1875.* Eds Robert Engberg and Donald Wesling. Salt Lake City, UT: The University of Utah Press.

Nacouzi, Salwa. 2005. Thomas Jefferson et les raisons d'une expédition : exploration, expansion, expansionnisme. In Caron, Nathalie & Naomie Wulf. *The Lewis and Clark Expedition.* Paris: Editions du temps.

Nott, Josiah. 1855. *Types of Mankind or Ethnological Researches Based Upon the Ancient Monuments, Paintings, Sculptures, and Crania of Races, and Upon Their Natural Geographical, Philological, and Biblical History.* Philadelphia: Lippincott, Grambo, and Co.

Noyes, John Humphrey. 1966. *Strange Cults and Utopias of 19th-Century America (Formerly titled: History of American Socialisms)* [1870]. New York: Dover.

O'Brien, Eugene. 2002. *Examining Irish Nationalism in the Context of Literature, Culture and Religion: a Study of the Epistemological Structure of Irish Nationalism.* Dublin: Edwin Mellen Press.

O'Cadhla, Stiofain. 2007. *Civilizing Ireland: O. S. 1824-1842, Ethnography, Cartography, Translation.* Dublin: Irish Academic Press.

Oliphant, Laurence. 1880. *The Land of Gilead: With Excursions in the Lebanon.* London & Edinburgh: W. Blackwood & Sons.

Osborne, Michael A. 2000. Acclimatizing the World: A History of the Paradigmatic Colonial Science. In *Nature and Empire: Science and the Colonial Enterprise, Osiris. A Research Journal Devoted to the History of Science and its Cultural Influences,* ed. Roy MacLeod. Second Series. Vol.15.

Paine, Thomas. 1966. *Common Sense and Other Political Writings* [1775]. New York: Pyramid Books.

Palmer, Alison. 2000. *Colonial Genocide.* Adelaide: Crawford House.

Parkman, Francis. 1980. *The Jesuits in North America in the Seventeenth Century.* Williamstown, Mass.: Corner House.

Parrington, Vernon Louis. 1930. *Main Currents in American Thought.* New York: Harcourt, Brace & World.

Patrick, Andrew William. 1857. *The Euphrates Railway: the Shortest Route to India*, London: Effingham Wilson.

Peterson, Merrill D. 1970. *Thomas Jefferson and The New Nation–A Biography*. New York: Oxford University Press.

Phillips, Ruth B. 2006. Show Times: de-celebrating the Canadian nation, de-colonising the Canadian museum, 1967-92. In *Rethinking Settler Colonialism: History and Memory in Australia, Canada, Aotearoa New Zealand and South Africa*. Ed. Annie E. Coombes. Manchester: Manchester U.P., 121-39.

Pinchot, Gifford. 1987. *Breaking New Ground* [1947]. Covelo, California: Island Press.

—. 2004. The Fight for Conservation (1910). In *Conservation in the Progressive Era: Classic Texts*. Seattle: University of Washington Press.

Porter, Roy. 2003. *Blood and Guts: A Short History of Medicine*. New York: Norton.

Powers, Henry. September 1898. The War as a Suggestion of Manifest Destiny. In *Annals of the American Academy of Political and Social Sciences.*

Pratt, Marie-Louise. 1992. *Imperial Eyes. Travel-Writing and Transculturation*. London, New York: Routledge.

Regard, Frédéric Ed. 2007. Introduction. *De Drake à Chatwin. Rhétoriques de la découverte*. ENS édition.

Regnauld, Henri. 1998. *L'Espace, une vue de l'esprit?* Rennes: Presses Universitaires de Rennes.

Richards, Thomas. 1993. *The Imperial Archive: Knowledge and the Fantasy of Empire*. London: Verso.

Rivière, J.L. and M. Llopès. 1980. *Cartes et figures de la terre*. Paris: Centre George Pompidou.

Robson, Charles. 1930. *The Influence of German Thought on Political Theory in the United States in the Nineteenth Century*. Ph.D. History department, University of North Carolina.

Ronda, James P. 2002. *Lewis and Clark among the Indians*. Lincoln, NE: University of Nebraska Press.

Rowley, C.D. 1974. *The Destruction of Aboriginal Society*. Ringwood (Vic.): Penguin Books.

Rudwick, Martin. 1976. *The Meaning of Fossils. Episodes in the History of Palaeontology*. New York: Neale Watson Academic Publications.

—. 1992. *Scenes from Deep Time. Early Pictorial Representations of the Prehistoric World*. Chicago and London: University of Chicago Press

Ruiz, Jean-Marie. 2001. Publius et la nature humaine. *Revue Française d'Etudes Américaines.* N°87.

—. 1999. Idéologie et tradition chez Mahan. *L'évolution de la pensée navale,* ed. Hervé Couteau-Bégarie. Vol.7. Paris : Economica.

Rydell, Robert W. 1999. 'Darkest Africa' African Shows at America's World's Fairs 1893-1940. In *Africans on Stage: Studies in Ethnological Show Business.* Ed. Bernth Lindfors. Bloomington (Indiana): Indiana U.P., 135-155.

Said, Edward W. 1979. *Orientalism.* New York: Vintage Books.

—. 1993. *Culture and Imperialism.* London: Chatto and Windus.

St. Jean de Brebeuf. 1957. *Les Relations de ce qui s'est passé au Pays des Hurons (1635-1648).* Genève: Droz.

Samuels, Ernest ed. 1973. *The Education of Henry Adams.* Boston: Houghton Mifflin.

Sarich, Vincent, and Frank Miele. 2004. *Race: The Reality of Human Differences.* Boulder: Westview Press.

Scheper-Hughes, Nancy. 2002. Coming to our senses: Anthropology and Genocide. In *Annihilating Difference: The Anthropology of Genocide.* Ed. Alexander Laban Hinton. Berkeley: U of California Press, 348-381.

Shelton, Anthony Alan. 2000. Museum Ethnography: An Imperial Science. In *Cultural Encounters: Representing Otherness.* Ed. Elizabeth Hallam & Brian Street. London & NY: Routledge, 156-194.

Schmitt, Peter J. 1990. *Back to Nature–The Arcadian Myth in Urban America.* Baltimore: The Johns Hopkins University Press.

Sigrist, René. 2008. "La République des Sciences" : essai d'analyse sémantique. In *La République des Sciences. Réseaux des correspondances, des académies et des livres scientifiques.* Dix-huitième siècle, 40.

Smith, Henry Nash. 1950. *Virgin Land–The American West as Symbol and Myth.* New York: Vintage Books.

Smith, Michael L. 1987. *Pacific Visions: Scientists and the Environment 1850-1915.* New Haven: Yale University Press.

Simpson, Moira. 1996. *Making Representations: Museums in the Post-Colonial Era.* London: Routledge.

Sinnett, Frederick. 1997. The Fiction Field of Australia (1871). In *The Oxford Book of Australian Essays* ed. Salusinszky, Imre. Oxford, Melbourne: Oxford University Press Australia.

Smyth, Gerry. 2001. *Space and the Irish Cultural Imagination.* UK: Palgrave Macmillan.

Smyth, William J. 2006. *Map-making, Landscape and Memory, a Geography of Colonial and Early Modern Ireland c.1530-1750.* Cork: Cork University Press, in association with Field Days.

Spengler, Oswald. 1952. *The Decline of the West.* Charles Scribners' Sons.

Stander, Jack. 1993. The 'Relevance' of Anthropology to Colonialism and Imperialism. In *The 'Racial' Economy of Science: Toward a Democratic Future.* Ed. Sandra Harding. Bloomington (Ind): Indiana U.P., 408-427.

Stanton, William. 1960. *The Leopard's Spots: Scientific Attitudes toward Race in America 1815-59.* Chicago: The University of Chicago Press.

Stepan, Nancy. 1982. *The Idea of Race in Science: Great Britain 1800 - 1960.* Macmillan Press.

Stephanson, Anders. 1999. *Manifest Destiny. American Expansion and the Empire of Right.* New York: Hill and Wang.

Street, Brian V. 1975. *The Savage in Literature, Representations of 'Primitive' Society in English Fiction 1858-1920.* London: Routledge.

Suberchicot, Alain. 2003. *Littérature américaine et écologie.* Paris : L'Harmattan.

Sumner, William Graham. 1934. *Essays of William Graham Sumner.* New Haven: Yale University Press.

Tabachnick, Stephen Ely. 1981. *Charles Doughty.* Boston: Twayne Publishers.

Taverdet, Gérard. 2007. (foreword to) *Espace Représenté, Espace Dénommé, Géographie, Cartographie, Toponymie.* Valenciennes: Presses Universitaires de Valenciennes.

Taylor, Alan. 2005. Jefferson's Pacific: The Science of Distant Empire, 1768-1811. In *Across the Continent: Jefferson, Lewis and Clark, and the Making of America.* Ed. Seefedt, Douglas, Jeffrey L. Hantman & Peter Onuf. Charlottesville: University of Virginia Press.

Taylor, Andrew. 1999. *God's Fugitive.* London: Harper Collins.

Taylor, Ian. 2003. *In the Minds of Men: Darwin and the New World Order.* Toronto: TFE Publishing. (5th edition).

Thoreau, Henry David. 1973. *The Maine Woods* [1864]. Princeton, New Jersey: Princeton University Press.

—. 1992. *Walden and Resistance to Civil Government* [1854]. Ed. William Rossi. New York, London: W.W. Norton & Company.

Thwaites, Reuben Gold. 1896-1901. *The Jesuit Relations and Allied Documents: Travels and Explorations of the Jesuit Missionaries.* Cleveland, OH.: Burrows Brothers Co., 73 volumes.

Tiffin, Chris and Alan Lawson. 1994. *De-Scribing Empire. Post-colonialism and Textuality.* London: Routledge.

Todd, Emmanuel. 1990. *L'Invention de l'Europe*. Paris: Seuil.

Tucker, Robert W. & David C. Hendrickson. 1990. *Empire of Liberty. The Statecraft of Thomas Jefferson*. New York: Oxford University Press.

Vincent, Bernard. 1999. *La destinée manifeste. Aspects idéologiques et politiques de l'expansionnisme américain au dix-neuvième siècle*. Paris: Ed. Messene.

Wale, Burlington B. 1893. *The Day of Preparation; Or the Gathering of the Hosts to Armageddon: A Book for the Times*, London: E. Stock.

Walls, Laura Dassow. 1995. *Seeing New Worlds: H.D. Thoreau and Nineteenth-Century Natural Science*. Madison: The University of Wisconsin Press.

Warren, Charles. 1875. *The Land of Promise, or Turkey's Guarantee*. London: George Bell & Sons.

Warren, Leonard. 1998. *Joseph Leidy: The Last Man Who Knew Everything*. New Haven: Yale University Press.

Weinberg, Albert. 1963. *Manifest Destiny*. Chicago: Quadrangle Books.

West, Shearer (ed). 1996. *The Victorians and Race*. Aldershot: Scolar Press.

Wilson, Margaret Oliphant. 1976. *Memoir of the Life of Laurence Oliphant and of Alice Oliphant, His Wife*. New York: Ayer Publishing.

Whelan, Kevin. 1992. Beyond a Paper Landscape – John Andrews and Irish Historical Geography. In *Dublin City and County: From Prehistory to Present*. Dublin: Geographical Publications.

Whitaker, Arthur P. 1954. *The Western Hemisphere Idea*. New York: Ithaca.

White, Morton and Lucia. 1977. *The Intellectual Versus The City–From Thomas Jefferson to Frank Lloyd Wright*. New York: Oxford University Press.

White, Richard. 1999. *The Middle Ground: Indians, Empires, and Republics in the Great Lakes Region, 1650-1815*. Cambridge University Press.

Wood, Denis. 1992. *The Power of Maps*. New York: Guilford.

Worster, Donald. 2002. *A River Running West: The Life of John Wesley Powell*. New York: Oxford University Press.

Young, Robert. 1995. *Colonial Desire: Hybridity in Theory, Culture and Race*. London: Routledge.

CONTRIBUTORS

Mehdi Achouche is a professeur agrégé who teaches in the Department of English at Jean Moulin University in Lyon. He is preparing a PhD on the representations of technological utopianism in contemporary Hollywood cinema, and researches more generally on the manifestations and role of the technological imagination in American history and culture.

Susanne Berthier-Foglar is a professor of American civilisation and Native American Studies at the Université de Savoie, in France. Her publications concern Native American history and ethnic identities in the Southwest from the first contact with Europeans to contemporary indigenous activism. She co-edited a book on *Sites of Resistance* (Paris: Editions Le Manuscrit, 2006), edited a book on the French in the Americas (*La France en Amérique*, Chambéry: Editions du LLS, 2009) and her book on Pueblo history has just been published by the University Press of Bordeaux (2010).

Sheila Collingwood-Whittick is a senior lecturer in the English Department of Stendhal University, Grenoble 3. Over the last thirty years her field of research has been that of postcolonial literatures and she has published widely on fictional and autobiographical writings from several former British colonies. More recently, her work, which has focused increasingly on Indigenous and non-Indigenous Australian fiction, has resulted in a collection of essays entitled *The Pain of Unbelonging: Alienation and Identity in Australasian Literature* (Rodopi, 2007) and several essays on the relationship between history and fiction in Australia. The scope of her research also encompasses non-literary issues; forthcoming publications include book chapters on the historical (in)visibility of Aboriginal Australians and Australia's continuing ambivalence *vis-à-vis* international legislation on the rights of Indigenous peoples, as well as a co-edited book on Indigenous peoples and genetic research.

Jean-Daniel Collomb has just been appointed senior lecturer in American civilisation at Jean Moulin University in Lyon. His PhD dissertation centered on the life and work of John Muir, the pioneering American environmentalist. He has published several articles on the writings of John Muir and his efforts to promote conservation in 19th and 20th century America.

Catherine Delmas is a professor of British literature at Stendhal University-Grenoble 3. Her PhD and field of research are about Western modes of representation of the Orient in 19th and 20th century fiction and travelbooks, as well as orientalist and (anti)imperialist discourse in the colonial and postcolonial eras. She has published several papers on Joseph Conrad, T.E. Lawrence, Rudyard Kipling, E.M.Forster, Lawrence Durrell, Michael Ondaatje, J.M. Coetzee in various periodicals and collective works. She published a book on *Ecritures du désert : voyageurs et romanciers anglophones XIXè XXè siècles* (*Writing the Desert from Burton to Ondaatje*) Presses Universitaires de Provence, 2005, and co-edited a book on *History-Stories of India* with professor Chitra Krishnan, University of Madras, India. (Delhi: Macmillan India, 2008).

Frederic Dorel is a senior lecturer and the head of the Department of Humanities at the Ecole Centrale in Nantes, France. His main research is on Native American Catholicism in the USA, on which he has published several papers. He has also contributed to several publications on US literature and history, as well as on the book industry in France and in Canada.

Anne Le Guellec-Minel is a senior lecturer in English at the University of Brest. She is a former student of the Ecole Normale Supérieure. Her Ph.D at the University of Nanterre (Paris) was on Patrick White's epic novels. Her research interests are in the area of Australian and Aboriginal Studies and more generally postcolonial studies and theory. She has published articles on representations of national identity in novels by Patrick White, David Malouf, David Foster, Richard Flanagan, but also by Aboriginal writers like Kim Scott and Bruce Pascoe.

Professor **Mark Meigs** teaches American Cultural History at the University of Paris-Diderot (Paris 7). His work concerns the cultural changes that took place in the United States at the end of the nineteenth and beginning of the twentieth centuries when the Americans started thinking of their culture as rich instead of needy. Dinosaurs bones number among their first important cultural exports. Professor Meigs is currently working on a cultural geography of Philadelphia showing the changes in importance attributed to different kinds of collections as they moved about that city.

Valérie Morisson has just been appointed senior lecturer at the university of Dijon. Her main research is on British civilization and aesthetics.

Stéphanie Prévost is a postgraduate researcher at François-Rabelais University in Tours, France, where she also currently teaches. She is completing her PhD on the reception of the Eastern Question in Britain (1875-1898). She recently published 'The Eastern Question and Britain's Foreign Policy (1876-1896)' (in Trevor Harris (ed.), *Art, Politics and Society in Britain (1880-1914): Aspects of Modernity and Modernism*, Newcastle: CSP, 2009) in which she explores the impact of the crusading ideal on Britain's late-nineteenth-century handling of the Eastern Question.

Jean-Marie Ruiz is senior lecturer of American Studies at the Université de Savoie, France. His research deals with American political theory and foreign policy; he is the author of the recently published *Une tradition transatlantique: L'impact du réalisme politique sur la fondation des Etats-Unis et la pensée politique américaine au 19e siècle*, Editions de l'Université de Savoie, 2010.

Richard Somerset is a lecturer at Nancy 2 University. He is primarily interested in the concept of historicity in the nineteenth century as manifested in the fields of natural science, historiography and literature. He has published articles on French and British pre-Darwinian evolutionary thought, evolutionary popularisation in France and in Britain, the historiography of Thomas Carlyle and Jules Michelet, 'prehistoric fiction,' and the scientific link connecting progressive liberalism to imperialism. He has also published specific studies of a number of nineteenth-century authors, including Jane Austen, George Eliot, Honoré de Balzac, Ralph Waldo Emerson, William Morris, Arthur Conan Doyle and Bram Stoker.

Donna Spalding Andréolle is a professor of American studies at Stendhal University-Grenoble 3 (France) where she teaches American history courses as well as seminars on the popular culture. Her research centers on 'low-brow' cultural objects as sites of social commentary as well as on representations of scientific progress in science fiction novels and Hollywood productions since the mid-twentieth century. Some of her most recent publications include the article in this book; "Between a Rock and a

Hard Place: Spaces of Entrapment in Big Love"; "Men are from Mars, Women are from Venus? A Case Study of Some Radical Feminist Discourse in the 1970s." She is currently preparing a collection of articles with Véronique Molinari entitled *Women in the Sciences, 17th century to Present*.

Christine Vandamme is a senior lecturer at Stendhal University, Grenoble 3, where she teaches British literature in the pre-modernist and modernist periods as well as postcolonial literature, notably that of Australia. She has published extensively on the representation of imperial space and its narratological as well as ideological and ethical implications, in Conrad's works but also in Patrick White's and David Malouf's fiction.

INDEX